MATHEMATICA®
For Calculus-Based Physics

Marvin L. De Jong
School of the Ozarks

Addison-Wesley is an imprint
of Addison Wesley Longman

Menlo Park, California • Reading, Massachusetts • Harlow, England •
Don Mills, Ontario • Sydney • Mexico City • Madrid • Amsterdam

Library of Congress Cataloging-in-Publication Data

De Jong, Marvin L.
    Mathematica® for calculus-based physics / Marvin L. De Jong
      p.   cm.
    ISBN 0-201-60339-X
    1. Mathematical physics. 2. Calculus. I. Title.
QC20.D33   1998
530.15'5—dc21

Mathematica® is a registered trademark of Wolfram Research, Inc.

Copyright © 1999, by Addison Wesley Longman.
Published by Addison Wesley Longman. All rights reserved.
No part of this publication may be reproduced, stored in a Retrieval system, or transmitted, in any form or by any means, electronic, mechanical, photocopying, recording, or otherwise, without the prior written permission of the publisher. Printed in the United States.

ISBN: 0-201-60339-X

1 2 3 4 5 6 7 8 9 10–COS–03 02 01 00 99

*Dedicated to my colleagues, past and present,
in the Mathematics-Physics department.
They have made my career joyful.*

Preface

To the Student

I am a constructivist. I believe that we construct our own understanding of the world in our own minds. That requires thinking rather than mindlessly following and memorizing instructions. We cannot be forced to think, we can only choose to think, and most of us resort to thinking only after every other option has been exhausted, including kicking and screaming. Many of my students are happy when they get the right answer regardless of how they got there, and they are anxious to go to the next problem. However, getting the right answer to the many problems in this book is not nearly as important as reflecting and thinking about how we got there and what we have learned. The process is every bit as important as the product. (Note: A constructivist is not a relativist; that is, a constructivist does not necessarily believe that the world view of a 10-year old is every bit as valid and useful as that of a Nobel prize-winning physicist.)

To the extent that you think about what you are doing, mastery of this book will give you a significantly deeper understanding of physics and will introduce you to a powerful computer algebra system, Mathematica. The ability of Mathematica to take the drudgery out of many calculations and its exciting graphics capabilities makes physics more enjoyable and understandable. Another payoff is the skill acquired in using a system (Mathematica) that is widely used in industry. The cost is learning and remembering a new language and its syntax. My hope is that you have some fun doing the problems and learning what Mathematica can do. I also hope that you learn to think about physics and acquire some power over Mathematica. It's up to you.

To the Instructor

To begin, this book is intended for students taking (or have taken) a calculus-based general physics course, or for somewhat more advanced students. Let's say the target audience is physics students in their first two years of study. Students in a non-calculus course will have a very difficult time.

Physics, mathematics, and Mathematica are all challenging if not downright difficult. So why combine all three in one book? A physics major or an engineering major will very likely take mathematics courses all through their undergraduate career. On the other hand, because in the world outside of education modern computer algebra systems are a key productivity tool, they are sometimes expected to learn a system such as Mathematica in a relatively short period, hopefully before they graduate. However, the learning curve for Mathematica is almost as long and steep as it is for mathematics. Our suggestion is to begin to learn Mathematica as early as possible, and to learn it in parallel with physics and mathematics. It is easier and more meaningful to the student that way. In our opinion, it is a wise college or university that exposes the same computer algebra system to its students in many of its science and mathematics courses starting early in the college experience.

What does Mathematica do for the student? It excites, engages, and empowers. Even learning to do such apparently trivial functions as solving equations and graphing is helpful. More than that, it can open up new possibilities such as studying the motion of objects in situations where normal analysis techniques fail. Mathematica puts powerful symbolic analysis, numerical analysis, and visualization techniques in the hands of students. Furthermore, it demands an active response rather than a passive one; Mathematica engages students.

On the other hand, we must be honest and point out that Mathematica can be very frustrating to students. If the computer's memory is inadequate or not managed well, Mathematica runs out of memory, the kernal disappears, the student is left studying a changeless screen, and all of the students' work is wasted when the computer is restarted. Students also seem to think that Mathematica should understand the "minor" changes in syntax that they continually make. Perhaps Mathematica should do this, but it doesn't, and the result is frustration. Then there are the incomprehensible error messages that mean nothing to the beginner. Finally, there are some calculations which take a lot of time; another source of frustration when the student has little idea what is happening. The solution is to have a laboratory with a Mathematica expert willing and available to help.

How should this book be used? We have a two-hour computational laboratory each week in our first year calculus-based physics course. In that lab we work with spreadsheets and Mathematica, in recent years mostly the latter. That is certainly one way to use the book. It can also be used as a workbook to be used in a computer lab outside of any regular class, or it can be used for independent study by a student. In any case, it is important that someone with considerable expertise in both physics and Mathematica be available to help the students at their computers. In addition, the instructor is responsible for selecting those problems and explorations that the students must do. Our recommendation is to do almost all of the problems and relatively few of the explorations, which take significantly more time and independent work. The problems should be worked in the order in which they occur in the text; that is very important. Do not delay the problems until the end of the chapter.

Some of the problems require execution of one or more Mathematica commands, others problems involve only physics, some problems are strictly mathematical, and many involve at least two of the skills: physics, mathematics, Mathematica. If you are going to do physics with Mathematica, then you must be able to do Mathematica and mathematics as well. We have not tried to constrain ourselves to a particular kind of problem.

One possible approach to working through the book is to have each chapter notebook on the computer and the student works through the notebook, executing the commands which in this case require no typing. The other approach is to provide the student with a printed version of the notebooks, and the student enters each command from the keyboard. We prefer the latter approach.

At the end of each chapter we have included a very brief review which we call **Mostly Mathematica**. Repetition is very important when learning Mathematica, and our experience suggests there cannot be to much of it. We urge you to ask your students to do this short set of exercises. We have also included some more extensive **Explorations** which students can do on their own or in groups. Theseproblems are generally more difficult, more time-consuming, and also more open-ended.

After working with assembly language, BASIC, and FORTRAN for many many years, it is difficult not to, with metaphoric hammer and anvil, forge Mathematica into the shape of those languages, DO loops and FOR ... NEXT ... loops included. We have tried very hard not to do that. There is not a single DO loop or FOR... NEXT... in the entire book (I think). We have tried to use the constructs that are characteristic of Mathematica. Our work represents an honest approach to Mathematica, not a hybrid of all the computer languages we have ever known. On the other hand, we have made no attempt to provide an exhaustive (and exhausting) exposure to Mathematica. We have used those commands which will be adequate for the problem at hand; no more, no less.

Perhaps the most controversial aspect of the book, if there is any, is the use of Mathematica's **DSolve** and **NDSolve** commands, starting in Chapter 7, to solve differential equations. Usually the first-year student is spared from differential equations until the chapter on oscillations, and then the student is provided with the solution. We should realize however, that every Newton's second law problem involves a differential equation. The only hurdle is to get students to write Newton's second law with the derivatives in place, rather than as $F = ma$. Students with a calculus background are be able to do that. Like many other things, it is a matter of exposure and practice. The numerical procedures in **NDSolve** are so powerful that it is a shame to constantly use Euler's method or some more sophisticated technique instead of **NDSolve**.

This book is challenging. The rewards to the student of its mastery are commensurate with the challenge.

Contents

Chapter 1 Calculations, Formulas, and Equations: A Mathematica Sampler

    1.1 Introduction 1

    1.2 Arithmetic in Mathematica 2

    1.3 Some Symbolic Computations 4

    1.4 Exact and Approximate Numbers 6

    1.5 The Internal Functions of Mathematica 7

    1.6 Calculus Operations 8

    1.7 Solving Simiple Problems 11

Chapter 2 Functions, Derivatives, and Numerical Integration

    2.1 Introduction 16

    2.2 Functions in Mathematica 17

    2.3 About Derivatives and Numerical Integration 21

    2.4 Quick Summary 30

    2.5 About our Programming Methods 31

    2.6 A Final Illustration of this Technique 33

Chapter 3 Raindrops, Pebbles, and Shuttlecocks: Objects Falling in Air

    3.1 Introduction 37

    3.2 On Raindrops and Pebbles 39

    3.3 Throwing Rocks 42

    3.4 Testing the Model 44

Chapter 4 Vectors, Baseballs, Planets, and Moonshots

4.1 Introduction 48

4.2 Simple Vector Operations in Two Dimensions 49

4.3 The Dot Product 50

4.4 Projectile Motion (in a vacuum) 51

4.5 A Vector Approach to the Euler-Cromer Method 53

4.6 Projectile Motion with Air Resistance 56

4.7 Orbital Motion 58

4.8 Moon Shots 62

Chapter 5 Using Mathematica to do Traditional Physics Problems

5.1 Introduction 68

5.2 Kinematics Problems 69

5.3 Statics Problems 74

5.4 Dynamics Problems 76

5.5 Energy Problems 78

5.6 Momentum Problems in One Dimension 81

Chapter 6 Potential Energy and Conservative Forces

6.1 Introduction 85

6.2 Potential Energy and Force 86

6.3 One-Dimensional Single-Particle Systems 88

6.4 Extensions to Two Dimensions 92

Chapter 7 Newton's Second Law is a Differential Equation

7.1 Introduction 100

7.2 Starting Simple: A Ball Thrown Upward 101

7.3 Objects Falling in a Resistive Medium 104

7.4 A Problem in Two Dimensions 107

7.5 Numerical Solutions of Differential Equations 109

Chapter 8 Topics in Linear Momentum and Gravitation

8.1 Introduction 117

8.2 Linear Momentum 118

8.3 Motion Under the Influence of Gravity 122

8.4 An Example from Biology 130

8.5 Swimsuit Section 133

Chapter 9 Oscillatory Motion

9.1 Introduction 134

9.2 The Simple Harmonic Oscillator 135

9.3 The Duffing Oscillator 137

9.4 Damped Oscillations 138

9.5 Damped and Driven Oscillations 140

9.6 Tracking the Duffing Oscillator on the Road to Chaos 141

9.7 The Lorenz Equations 145

Chapter 10 Topics in Wave Motion

10.1 Introduction 148

10.2 The Mathematica Expression of a Wave 149

10.3 Beats: A Simple Example of Interference 151

10.4 Interference: A Mathematica Approach 153

10.5 Examples of Interference 156

10.6 Diffraction and Interference 163

10.7 The Wave Equation 165

Chapter 11 Electric Potential Problems

11.1 Introduction 170

11.2 One-Dimensional Potentials 171

11.3 Electric Potential in Two Dimensions 174

11.4 Three-Dimensional Potential Functions 178

11.5 The Uniformly Charged Sphere 179

Chapter 12 Electrical Circuits

12.1 Introduction 182

12.2 dc Circuits: Kirchoff's Law Problems 183

12.3 The *RC* Circuit 184

12.4 The *LRC* Circuit 188

12.5 The *LR* Circuit 194

Chapter 13 Return to Chaos

13.1 Introduction 197

13.2 The Quadratic Map: A Simple Approach 198

13.3 The Quadratic Map: Part II 200

13.4 Return of the Duffing Oscillator 204

13.5 What is Chaos? 212

Chapter 14 Mathematica in the Laboratory

  14.1 Introduction 217

  14.2 Graphing and Analyzing Data 219

  14.3 More Complex Data 228

  14.4 Importing Data 231

Chapter 15 Additional Topics on Waves

  15.1 Introduction 237

  15.2 Fourier Synthesis and Analysis 238

  15.3 Discrete Fourier Analysis 243

  15.4 A Quick Look at Quantum Mechanics 249

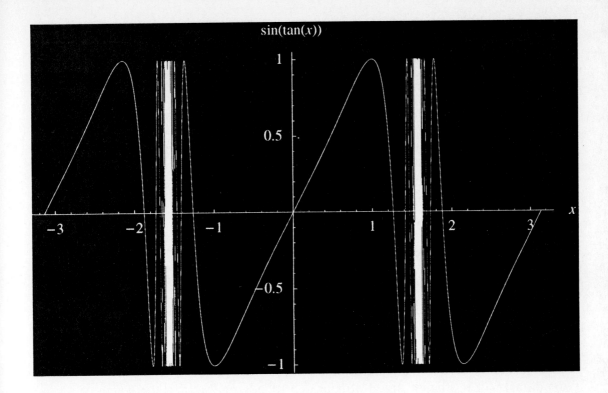

CHAPTER 1 Calculations, Formulas, and Equations: A Mathematica Sampler

■ 1.1 Introduction

Welcome to doing physics with Mathematica. With its computational power and beautiful graphics, Mathematica can help you learn physics and make physics more enjoyable, absorbing much of the drudgery. It can, for example, perform differentiations, integrate, and be used to solve sets of equations, a mathematical task that is a ubiquitous source of error for students. In this chapter our intention is to give you some idea of the scope of Mathematica. Do not expect to be able understand or memorize all of the syntax you see in this chapter; that takes a great deal of time and practice. This chapter is intended to be a tasting experience, not a meal.

■ 1.2 Arithmetic in Mathematica

You should be able to locate the Mathematica icon on your computer monitor screen and get Mathematica started on your computer. When you do, you will be in a Mathematica notebook, and you can simply begin to type an instruction. Mathematica will print the instruction in boldface type. Once you type the instruction, press the ENTER key (not the RETURN key). Try the following instruction, which includes addition and subtraction:

> To execute the following command, type **6 + 8 – 2 + x**, then, with the cursor anywhere on the same line as the command, press the ENTER key. Do not type the **In[1]: =** part of the instruction.

In[1]:= **6 + 8 – 2 + x**

Now you know how to add and subtract with Mathematica.

We will generally use the SI system of units and most of our calculations will be accomplished without specifying units, in which case you may assume they are SI. When we choose another system of units, we will be careful to specify what they are.

A light-year is the distance light travels in a year. How far away in meters is a star 4 light-years distant? Let c be the speed of light, $3.0 \cdot 10^8$ m/s, and let t be the length of a year, $3.2 \cdot 10^7$ s. A space between two numbers or symbols is an assumed multiplication. To find the distance in meters we must compute the boldfaced command below,

> To execute the following command, put the cursor anywhere on the same line as the command, then press the ENTER key. We will no longer repeat this advice.

In[2]:= **4 3.0 10^8 3.2 10^7**

and the answer will be in meters. *Notice the spaces between each number to be multiplied* including a space between the number and its 10^n multiplier. Now you know how to multiply and do simple exponents in Mathematica.

Problem 1. To divide, you can either use a **/** or the divide selection from the palette, $\frac{\Box}{\Box}$. Here is a division problem. A gas at a pressure of 3.0 Pa has a volume of 4.2 m^3. What will be its pressure if the volume is reduced, at constant temperature, to 2.3 m^3? Since

$$P_2 = P_1 \frac{V_1}{V_2} \qquad (1)$$

this problem can be solved with a multiplication and a division. Try it in Mathematica. You should get 5.5 Pa. Rather than assign numbers to three variables and write the

1.2 Arithmetic in Mathematica

formula symbolically, simply find P_2 with a single calculation. At this point you should be able to add, subtract, multiply, and divide.

Problem 2. Mathematica does exponentiation with either the carrot symbol, ^, or by choosing $\square^\square$ from the palette and filling in the appropriate spots. Find: 2^3, 3^{-3}, 5^{20} and 7^{-50}.

> In our experience, exponentiation is done more easily with the carrot symbol ^ than with the palette. Palettes become more useful with more complex operations such as integrals.

Problem 3. Mathematica does fractional exponentiation with fractional exponents (the carrot symbol), $\sqrt{\square}$ from the palette, $\sqrt[\square]{\square}$ from the palette, or $\square^\square$ from the palette. For example:

In[3]:= **27^(1/3)**

and

In[4]:= $\sqrt[3]{27}$

will both give the same answer. What does

In[5]:= **27^1/3**

give? Try these calculations: $3^{1/2}$, $10^{-1/3}$, $4^{1/4}$, $(\frac{3}{5})^{-3}$.

At this point, you should perceive a significant difference between your calculator and Mathematica. Your calculator would crank out decimal approximations to a calculation such as $\sqrt{3}$, while Mathematica does not. We will discuss that in more detail in a moment.

Mathematica can also do complex arithmetic, but it represents $\sqrt{-1}$ by **I**, not *i*. Find:

In[6]:= **I^2**
       (1 + I)^2
       Abs[3 + 4 I]
       $\dfrac{1 - I}{1 + I}$
       (a − b I)^3

> When doing complex arithmetic, you still need a space between a symbolic multiplier of **I** and **I** itself, for example, **a** times **I** is **a I**.

1.3 Symbolic Computations

Which exerts the larger gravitational force on the Moon, the Earth or the Sun? According to Newton, the gravitational force between masses M and m is

$$F = G\frac{Mm}{r^2} \qquad (2)$$

In Mathematica we may write the same thing in almost the same way:

In[11]:= $F := \dfrac{G\,M\,m}{r^2}$

> The use of := rather than simply = to make an assignment may seem peculiar. Please accept it for now; we will explain it later.

For the Sun-Moon interaction, we assign the following values in Mathematica:

In[12]:= $G = 6.67\ 10^{-11}$; $M = 1.99\ 10^{30}$; $m = 7.36\ 10^{22}$; $r = 1.50\ 10^{11}$;

so the force of the Sun on the Moon is

In[13]:= F

To find the force of the Earth on the Moon, we need to change M and r.

In[14]:= $M = 5.98\ 10^{24}$; $r = 3.84\ 10^{8}$;

In[15]:= F

Clearly the force of the Sun on the Moon is a little more than twice as large as that of the Earth on the Moon.

There is another, perhaps better, way to do this problem. First we clear all of the old variables so we can reuse them, an effort that is always a good idea before beginning a new problem in Mathematica. There are several ways to clear variables so they may be used again. We could write

In[16]:= **Clear[F, G, M, m, r]**

or, if we have used only lowercase variable, we can clear all of them with the following command

In[17]:= **Clear["@"]**

or, finally, we can clear all user-defined variables with a

1.3 Symbolic Computations

In[18]:= **Clear["Global`*"]**

command.

> The **Clear** command may be the *most important* command to remember. Failure to clear variables can produce many puzzling results and cause a lot of grief. If you are having any trouble at all that you do not understand, try clearing the variables.

We will repeat these calculations, but this time we will have different names for the gravitational forces of the Sun and Earth on the Moon, as well as different names for the masses of the Sun and Earth. For the Sun,

In[19]:= $F_s = \dfrac{G M_s m}{r_s^2}$

while for the Earth,

In[20]:= $F_e = \dfrac{G M_e m}{r_e^2}$

> Use the palette entry □□ to enter subscripts.

Because they cancel in the ratio, clearly we do not need G or the mass of the Moon. But we do need the following quantities:

In[21]:= $M_s = 1.99 \; 10^{30}; \; r_s = 1.50 \; 10^{11}; \; M_e = 5.98 \; 10^{24}; \; r_e = 3.84 \; 10^{8};$

and now we ask for

In[22]:= $\dfrac{F_s}{F_e}$

Problem 4. What effect does a semicolon after an assignment or calculation have on the output?

Problem 5. Express distance as the product of speed and time. Assume that the speed is 50 m/s. Use Mathematica to find the distance traveled in 2.3 hours.

Problem 6. The general gas law is $PV = nRT$, where R is the gas constant. How will the pressure P of a gas change if its volume V is reduced to $\frac{1}{3}$ of its original value while its temperature T is doubled? This time, do not use regular subscripts and begin by defining **P1** as

In[23]:= **P1 = n R T1 / V1**

Then define an appropriate **P2**, and calculate the ratio $\frac{P1}{P2}$.

One of the most common errors in working with Mathematica is to not space variables you intend to multiply. For example, suppose that in the previous calculation we had written

In[24]:= **P1 = nRT1 / V1**

then the computer would assume that **nRT1** was a single variable. Don't forget spaces between variables you wish to be multiplied.

■ 1.4 Exact and Approximate Numbers

When supplied with exact numbers, Mathematica will always give exact answers unless you specifically ask it for a decimal approximation. What is an exact number? Some rational numbers can be represented exactly by decimals – for example, 1/4 = 0.25 – but most cannot – for example, 1/3 is only approximately 0.3333. In general, then, Mathematica will return a rational number answer as a rational number, not a decimal approximation. However, if you provide an approximate number (decimal) as input, you will get an approximate number as output. Type

In[25]:= $\dfrac{1}{4.}$

and you will get a decimal answer. In general, if a decimal point is included in a number in the calculation, then a decimal number will be returned. Of course, irrational numbers cannot be expressed exactly as decimals, so, for example,

In[26]:= **Sin$\left[\dfrac{\pi}{3}\right]$**

returns $\sqrt{3/2}$, while

In[27]:= **Sin$\left[\dfrac{\pi}{3.}\right]$**

returns 0.866025. If a decimal approximation is needed, use the **N** (numerical) function. For example, in the unlikely event we need the sine of $\pi/3$ to 12 significant figures and π to 100 significant figures, write

In[28]:= $\mathbf{N}\left[\mathbf{Sin}\left[\dfrac{\pi}{3}\right], 12\right]$

$\mathbf{N}[\pi, 100]$

A convenient way to get an approximate value is to use the *postfix* form of the **N** command

In[30]:= **Log[10] // N**

which gives the answer approximate to six significant figures. We will use the postfix form quite frequently in this book.

Problem 7. Find arcsin(1/2) exactly and to six significant figures using the **N** function. The syntax of arcsin is **ArcSin[***argument***]**, but you can look up the syntax of any command by prefacing it with a question mark, **?ArcSin**, for example.

Problem 8. Find **Log[π]** exactly and to 10 significant figures. Try to give the reason for Mathematica's exact answer. Is there an exact answer to $\log(\pi)$?

■ 1.5 The Internal Functions of Mathematica

Mathematica will have all the standard functions that you find on your calculator and many more. The important thing to know is the syntax. The arguments of functions are always expressed in square brackets. You will save yourself a lot of grief if you remember this fact and the fact that square brackets are never used in computations to indicate order of operations. Here are three examples of internal functions:

In[31]:= **Tan[π]**
Log[*e*]
Exp[x]

Square brackets are only used for functions. Here is a calculation that won't work because set brackets and square brackets are *never* used for order of operations in an algebraic calculation in Mathematica.

In[34]:= $5 \{[(x - 2)^2 + 3y](x - 2)\}$

To do this calculation correctly, write

In[34]:= $5\,(((x - 2)^2 + 3y)(x - 2))$

and if you want to expand it, write

In[35]:= **Expand**$[5\,(((x - 2)^2 + 3y)(x - 2))]$

Here are three examples of functions within functions. Execute them and try to understand what is happening.

In[36]:= $\text{Sin}[\text{ArcSin}[\frac{1}{2}]]$

$\text{ArcTan}[\text{Sin}[\frac{\pi}{6}]]$

$\text{Sin}[\text{ArcTan}[\text{Log}[\pi]]]$

Problem 9. Here are some Mathematica functions that we don't normally think of as functions. Execute them, then explain what each of them does.

In[39]:= $\text{Expand}[(x + y)^4]$

$\text{Factor}[x^4 - y^4]$

$\text{Solve}[a\,x^2 + b\,x + c == 0, x]$

$\text{Plot}[x^2 + 2, \{x, -2, 2\}];$

$\text{Plot}[\text{Tan}[\text{Sin}[x]], \{x, -\pi, \pi\}];$

$\text{Plot3D}[\text{Sin}[x\,y], \{x, 0, 2\pi\}, \{y, 0, \pi\}, \text{PlotPoints} \to 30];$

> Mathematica calls practically everything a function. We will frequently refer to these functions as commands or instructions, keeping the word *function* reserved for traditional functions in mathematics and physics. In the third function, above, note the use of ==. We explain this syntax in a subsequent paragraph.

Problem 10. Experiment with the command ? followed by a function. For example,

In[43]:= **? Plot**

What does the question mark do? Also try a double question mark, like this:

In[44]:= **?? Plot**

What is the difference between the **?** help command and the **??** help command?

■ 1.6 Calculus Operations

Recall the definitions of velocity and acceleration for motion in one dimension. The definitions can be either in the form of derivatives or integrals. We repeat them here for the sake of completeness. Here is the definition of velocity:

1.6 Calculus Operations

$$v = \frac{dx}{dt} \text{ or } x_t = \int_0^t v(t)\,dt + x_o \quad (3)$$

Here is the definition of acceleration:

$$a = \frac{dv}{dt} \text{ or } v_t = \int_0^t a(t)\,dt + v_o \quad (4)$$

> Mathematica's integrals use the differential symbol dt as opposed to the more traditional dt.

A more traditional way of writing the integral in Equation (3), for example, is

$$x_t - x_o = \int_0^t v(t)\,dt \quad (5)$$

However, Mathematica cannot use this kind of assignment notation with two variables on the left-hand side of the equal sign. Use Equation (3), not Equation (5).

1.6.1 Differentiation

In[45]:= **Clear["@"]**

Mathematica can differentiate and integrate. Differentiation can be done in several different ways, with the **D** operator, with the palette entry $\partial_\Box \Box$, and with prime notation. We will reserve the latter notation for the next chapter where we introduce function notation. Suppose the position of a particle is given by

In[46]:= $x = \frac{1}{2} g t^2$

The velocity is the time derivative of the position, so we can find v as follows:

In[47]:= $v = \partial_t \, x$

or with the **D** operator,

In[48]:= **D[x, t]**

Acceleration is the derivative of the velocity, and can be found by taking another derivative. There are two ways to find second derivatives:

In[49]:= $\partial_t \, v$

Or

In[50]:= $\partial_{t,t}$ x

Problem 11. A Miata is moving along the x-axis in such a way that its position

$$x = -(t-1)^2 + 3t \qquad (6)$$

Find its velocity and acceleration.

Problem 12. When is the velocity of the Miata zero? What is the maximum value of x for the Miata?

1.6.2 Integration

In[51]:= **Clear["@"]**

The acceleration of a high-speed passenger train is $a = 6t$ m/s/s. If it starts from rest, how fast will it be going in 5 s? Where will it be after 5 s, assuming it started at the origin? Here is one way to proceed.

In[52]:= **a = 6 t; v_0 = 0; x_0 = 0;**

Now we use the integral definitions of velocity and position:

In[53]:= $v_t = \int_0^t a\, dt + v_0$

$x_t = \int_0^t v_t\, dt + x_0$

from which we may calculate the velocity and acceleration quite easily. Both the velocity and the position are returned in terms of the time. How could we evaluate these integrals for $t = 5$? We append the symbolism /.t->5 to a command, which may be taken to mean *given that t is 5*, like this:

In[55]:= v_t /. t -> 5
x_t /. t -> 5

> The symbolism **t -> 5** is called a *rule* in Mathematica, and, as opposed to an *assignment*, such as **t = 5**, the rule *does not* result in the value of 5 being *permanently* assigned to t. Another way of thinking of what /.t -> 5 means is: *if t = 5*. With the expression for v_t above, we are finding the value of the integral if $t = 5$. This particular kind of rule, called a *replacement rule*, will be used often in this book.

Problem 13. A car is traveling at 48 m/s in the +x-direction at $t = 0$ when it brakes to a stop with a constant acceleration of $a = -t^2$ m/s/s. Begin by finding v at any time t with an integral. How fast is it going after 3 s? How long will it take to stop? How far does it go before stopping? (You need to do another integration.) Be sure to clear any variables you may choose that duplicate the variables previously used.

> Mathematica Version 3.0.0.0 does not allow us to clear variables with subscripts such as v_o with an ordinary **Clear** command such as we have been using. This is a big disadvantage to their use, and we will rarely subscript variables; we will write **vo** rather than v_o. You can clear real subscripted variables one at a time using a command similar to the one that follows.

In[57]:= **v₀ =.**

1.7 Solving Simple Problems

One way to improve your Mathematica skills is to practice doing simple problems whose solutions you can find with pencil and paper, or whose solutions are given in your physics textbook. Here we provide some typical examples.

Example Problem 1. A ball is thrown straight up into the air with an initial velocity of 50 m/s. How high will it go? The equation governing the motion of the ball is

$$y = y_0 + v_0 - \frac{1}{2} g t^2 \tag{7}$$

In Mathematica we write

In[58]:= **Clear["@"]**

In[59]:= **y = yo + vo t - $\frac{1}{2}$ g t²**

(Notice we are not using real subscripts.) We take the coordinate system origin at ground level. The various parameters are now

In[60]:= **yo = 0; vo = 50; g = 9.8**

Our strategy is to find when the velocity is zero, because that occurs at the top of the flight. We assign

In[61]:= **v = ∂$_t$ y**

Mathematica's **Solve** command will help us if we can't solve this equation mentally,

In[62]:= **Solve[v == 0, t]**

> Observe the use of the double equal, == in Mathematica's **Solve** command. Mathematica uses the single equal sign for *assignments*, **g = 9.8**, and the double equal for *conditional equations*. Human beings can distinguish between an assignment and a conditional equation from the context; however, Mathematica is not a human being.

Finally, we use a replacement rule to calculate y when t is 5.10204.

In[63]:= **y /. t -> 5.10204**

Example Problem 2. Consider the same problem, and find how long the ball is in the air and how fast it is going when it hits the ground. When it hits the ground we know that $y = 0$.

In[64]:= **Solve[y == 0, t]**

There are two solutions, the first corresponds to when the ball leaves the ground. The second is the one we want. To find v, we use a replacement rule and calculate

In[65]:= **v /. t -> 10.2041**

Problem 14. The bad guys leave the bank at $t = 0$ and travel at 10 m/s. The posse leaves the saloon next to the bank 50 s later, and their horses can travel at 12 m/s (at least for awhile). How long after the bad guys leave the bank do the good guys catch up to them? How far did the posse gallop to catch the bad guys?

> You can easily do this problem in your head or on a piece of paper. Using Mathematica to solve it is like hunting a mouse with an elephant gun. However, you will gain expertise with Mathematica by beginning with simple problems.

Problem 15. This problem is more difficult, but it's likely you have seen something similar in your calculus book. Starting at $t = 0$ a mouse runs along a meter stick on the x-axis. The position of the mouse is given by

$$x = \frac{t^3}{3} + \frac{t^2}{2} - 2t \tag{8}$$

At what times, if any, does the mouse change direction? (It will be helpful to find where the speed of the mouse is zero, because the mouse cannot change direction without its speed going to zero.) On what intervals for $t \geq 0$ is the mouse is moving right? Moving left? On what positive time intervals is the mouse is speeding up? Slowing down? To assist you, use the $\partial_t x$ command and the **Solve** command. You might even plot x versus t to see whether you are on the right track.

1.7 Solving Simple Problems

Having introduced the subject of solving equations, we point out that Mathematica can also solve equations symbolically. Suppose you have a formula such as

$$\frac{1}{r} = \frac{1}{x} + \frac{1}{y} + \frac{1}{z} \qquad (9)$$

and you wish to solve for z. You can use the **Solve** command this way:

In[66]:= **Clear["@"]**

In[67]:= **Solve$\left[\frac{1}{r} == \frac{1}{x} + \frac{1}{y} + \frac{1}{z}, z\right]$**

Example Problem 3. A 1-kg crow lands in the middle of two wires stretched between two powerline poles. If the wire stays essentially straight from the poles to the crow and each of the two segments makes an angle of 2 degrees with respect to the horizontal, what is the tension in the wire?

Begin by drawing a free-body diagram. We place the point upon which the crow sits in equilibrium. There are two tension forces, one upward to the left of the crow and one upward to the right. Resolving these forces into components gives two equations:

In[68]:= **Clear["Global`*"]**
        eq1 = T1 Cos[2 Degree] − T2 Cos[2 Degree] == 0
        eq2 = T1 Sin[2 Degree] + T2 Sin[2 Degree] − 19.8 == 0

We solve this set of simultaneous equations this way:

In[71]:= **Solve[{eq1, eq2}, {T1, T2}]**

Aha, the tensions are same, as expected, and each is approximately 140 N.

Example Problem 4. If the wire between the two utility poles in the previous problem is strung very tightly, then the angle with the horizontal becomes smaller. How does the tension in the wire change in this case? Here we might use a **Plot** instruction. To get an exact solution, we **Clear** the variables we used previously and use a symbolic g for the acceleration of gravity. Thus

In[72]:= **Clear["Global`*"]**
        eq1 = T1 Cos[θ] − T2 Cos[θ] == 0
        eq2 = T1 Sin[θ] + T2 Sin[θ] − 1 g == 0
        Solve[{eq1, eq2}, {T1, T2}]

> This is an example where we put all of the instructions in one cell to execute them with one stroke of the ENTER key. Cell marks are not shown in this book, but they will be shown on the right hand side of the page in your Mathematica on-screen notebook.

Next, we plot the tension as a function of θ, T1 = T2 = 4.9 Csc[θ], from 0.01 radian to 0.5 radian.

In[76]:= **Plot[4.9 Csc[θ], {θ, .01, 0.5}, AxesLabel −> {"θ", "T"}];**

and we see that the tension goes to infinity as the angle approaches zero. That is why wires between utility poles are not stretched too tightly. There is probably not much worry with crows and squirrels, but ice and wind loading are very significant and it pays to have some droop in the wires.

Problem 16. A 10-kg mass is hung from a pole by a wire. Another wire attached to the mass is used to pull the mass horizontally until the first wire makes an angle of 23 degrees with respect to the horizontal. Use Mathematica to find the tension in each wire. Also try to graph the tension in the horizontal wire as a function of the angle that the first wire makes with the horizontal.

Problem 17. A 15-kg mass is firmly attached to a 5-kg bathroom scale, which in turn is sliding down an inclined plane. The plane is inclined at 35 degrees and the coefficient of kinetic friction is 0.11. What does the scale read? What is the acceleration of the scale/block system? Be sure to draw a free-body diagram before beginning. Supposing the angle of the incline is variable, at what angle will the scale/block system slide down at constant velocity?

Mostly Mathematica

1. Find, to eight significant figures:

$$\frac{1}{2 \cdot 10^6} + \frac{5}{3 \cdot 10^7} \tag{10}$$

2. Solve for x, then for r:

$$z = -\frac{rxy}{rx + ry - xy} \tag{11}$$

3. If

$$y = 2x^2 - 4x + 1 \tag{12}$$

at what value of x is y a minimum? What is the area under the curve of y versus x on the interval [0, 1]? Make a graph of y versus x.

1.7 Solving Simple Problems

Exploration

Look up and experiment with some of the graphing *options* of the **Plot** instruction. You might try graphing multiple graphs with dotted and dashed lines, experimenting with color, and adding text to a graph for axis labels and a title. The command

In[77]:= **?? Plot**

lists the options. Options are given in the form of rules. Here is a **Plot** instruction.

In[78]:= **Plot[Sin[Tan[x]], {x, −π, π}];**

Only the essentials are given. Any other addition to this instruction in the form of rules are called *options*. Here is the **Plot** command used to generate a graph similar to the one at the head of the chapter. Notice the use of several options.

In[79]:= **Plot[Sin[Tan[x]], {x, −π, π},**
       PlotPoints −> 1000, Background −> GrayLevel[0],
       DefaultColor −> GrayLevel[1], PlotLabel −> "Sin(Tan(x))",
       AxesLabel −> {"x", ""}, TextStyle −> {FontFamily −> "Times",
       FontSlant −> "Italic", FontSize −> 14}, PlotStyle −> {Thickness[.0001]}];

Closing Comments

When you use Mathematica, avoid a lot of troubleshooting time and save yourself an enormous amount of grief by

    1. *Clearing variables between problems.*

    2. *Remembering that all Mathematica functions*

        A. *start with a capital letter and*

        B. *have their arguments enclosed in square brackets.*

    3. *Not using set brackets or square brackets to indicate order of operations.*

    4. *Making sure you have spaces between variable names when multiplication is intended.*

    5. *Using a question mark or double question mark prefix on a command to get syntatical help rather than guessing at syntax.*

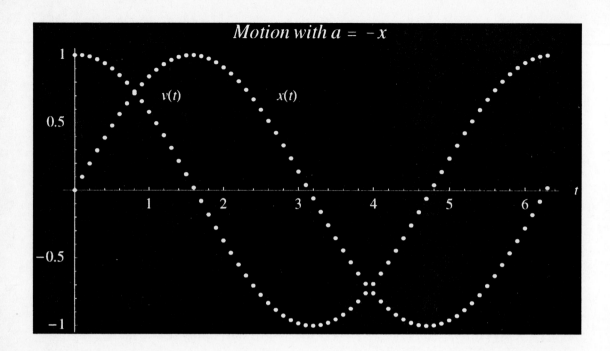

CHAPTER 2 Functions, Derivatives, and Numerical Integration

■ 2.1 Introduction

The principal goal of this chapter is to introduce a powerful but simple mathematical technique to study the motions of physical systems. This numerical technique makes use of the definition of the derivative of a function, and it allows us to predict the velocity v and position x of an object even when the forces acting on the object are rather complex, for example, the motion of a raindrop through the air. Of course, the basic postulate that lies at the foundation of classical mechanics and our work in this chapter is Newton's second law of motion, which in one dimension is

$$F = ma \qquad (1)$$

For our purposes in this chapter this postulate might better be written

$$\sum F_{\text{external}} = ma = m\frac{dv}{dt} = mv'(t) \qquad (2)$$

where the summation symbol emphasizes that we must find the *sum* of the external forces and the *derivative* means we cannot simply solve for v directly, instead we must integrate a to find v. Thus

$$v(t) = \int a(t)\, dt = \int \frac{1}{m} \sum F_{\text{external}}\, dt \qquad (3)$$

Furthermore, to find the position x we must integrate once more. Thus

$$x(t) = \int v(t)\, dt \qquad (4)$$

You may wish to begin by using your calculus book to refresh your memory of the definition of the derivative of a function. In order to exploit the technique developed in this chapter you will need a good understanding of the derivative, and before we begin we will need to learn how to express functions in Mathematica.

Problem 1. Students sometimes do not know what it means to *apply Newton's second law* to a problem, do you? You should be able to do this before you delve much more deeply in this book. Here is a problem to try: A raindrop of mass m falling in air experiences the downward force of gravity and an upward force of air resistance proportional to the velocity squared, $F = kv^2$. If we take positive as being down, apply Newton's second law to this problem and solve for the acceleration. Begin by drawing a free-body diagram.

> In this chapter and in all notebooks and chapters to follow, you will be expected to enter and execute the Mathematica commands in the sequence in which they are given. Failure to do so can produce error messages and sometimes rather meaningless results.

■ 2.2 Functions in Mathematica

Much of calculus and physics is about change. A neutrino leaves the interior of the Sun after a nuclear reaction and changes the interior of the sun forever, but not by much. Spring rains on the plains send floodwaters pouring into a local reservoir causing the water level to rise. A rocket lifts off the ground carrying the space shuttle. The planets change their positions in the sky from night to night. All of these changes may be described with functions and their derivatives, and that is why courses in calculus are a prerequiste for much of physics and for this book.

In calculus, functions are almost invariably described as *f(x)* = *expression*. In physics, we would like the notation to be slightly more meaningful, so we may choose symbols different than *f(x)*. For example, if a car is moving down the road, we might describe its position as a function of time by $x(t)$, its velocity by $v(t)$, and its acceleration by $a(t)$. In the case of an object thrown straight up into the air, we might locate the origin of our coordinate system on the ground and write

$$y(t) = y_0 + v_0 - \frac{1}{2} g t^2. \tag{5}$$

In Mathematica, the function notation is only slightly different than in calculus and physics; it uses square brackets. Using the same example we would write:

In[1]:= $y[t\_] := yo + vo\, t - \dfrac{g\, t^2}{2}$

> One of the most common errors in defining a function is to forget the blank character, _, on the left of the equal sign.

Now enter

In[2]:= $y[t]$

> Mathematica may reorder in its own fashion the terms you enter.

2.2.1 A Brief Aside About Mathematica's = and :=

Mathematica actually has three kinds of equal signs, and there are important but subtle differences between them. In Chapter 1, we examined the difference between an *assignment* and a *conditional equation*; in the latter case we use the double equal, ==. In this chapter we will look at the difference between = and :=, both of which are *assignments*. We will illustrate with an example. Define g on the Earth as

In[3]:= $g = 9.8$

and then define the distance an object falls from rest with the two different assignment equal signs, = and :=. First we use the = sign.

In[4]:= $s[t\_] = \dfrac{1}{2} g\, t^2$

Next, we use the := sign.

In[5]:= $h[t\_] := \dfrac{1}{2} g\, t^2$

From the graph you can get approximate values for the time spent in the air, the maximum height, and the range.

> On some versions of Mathematica, you can place the cursor on a point on the graph, hold down a control key, and read the coordinates of the point. Try it on your system.

Problem 5. Use the graphs you just made to find the approximate values for the time spent in the air, the maximum height, and the range.

Problem 6. There are three graphs on the coordinate system produced by the last instruction. Using the same graphs and your knowledge that the derivative gives the slope of the curve, how can you tell which is the position graph and which are the velocity and the acceleration graphs?

Problem 7. A policewoman waits at rest along the side of the road for an unsuspecting speeder who passes her going 80 mph (66 ft/s). If the policewoman starts the chase the very moment the speeder passes her, at what distance will she overtake the speeder if her car can accelerate at a constant 5 ft/s/s? Define two functions, one for the speeder's position and one for the policewoman's position. Graph them versus time to find the approximate solution to the problem.

Problem 8. Two monkeys hang from a branch. One monkey drops and 1 s later the other monkey drops. Does the distance between the monkeys increase, decrease, or remain the same once both monkeys are dropping? Use Mathematica to graph $y(t)$ for each monkey on the same coordinate system. Then answer the question. Use the same graph to decide if they hit the ground 1 s apart, more than 1 s, or less than 1 s. This problem is an adaptation of one in Epstein (1983).

■ 2.3 About Derivatives and Numerical Integration

Derivatives measure the rates of change of functions and because calculus and physics are about change, derivatives are widely used in physics, chemistry, and engineering. In calculus you encountered the definition of the derivative, which will be be very important for our work. Here it is in the unlikely case you have forgotten it:

$$f'(x) = \lim_{h \to 0} \frac{f(x+h) - f(x)}{h} \qquad (6)$$

Although we will generally use the prime notation, many times scientists prefer the following notation for the derivative:

$$\frac{d}{dx} f(x) = \lim_{h \to 0} \frac{f(x+h) - f(x)}{h} \qquad (7)$$

In physics, the functional notation $f(x)$ is often changed to give a clearer idea of what quantity is varying. For example, a car travels down a level road along which an x-axis is conveniently located, and at time t its position is $x(t)$. Thus, $x(t)$ replaces $f(x)$ as the function in which we are interested and the derivative becomes.

$$v(t) = x'(t) = \lim_{h \to 0} \frac{x(t+h) - x(t)}{h} \tag{8}$$

If h is *small enough* but not zero, then

$$v(t) \simeq \frac{x(t+h) - x(t)}{h} \tag{9}$$

and we may take some liberty and rearrange this expression to give us the approximation

$$x(t+h) \simeq x(t) + v(t)h. \tag{10}$$

This is a very important approximation, and we shall use one form or another of it many times to find approximate values for positions of things that are accelerating as a result of external forces. This approximation becomes an exact equality only if the slope of the $x(t)$ versus t curve is a straight line, that is, if $v(t)$ is a constant, but it is a very useful approximation in many situations. It says simply that the position, $x(t+h)$, at the end of some time interval h is the position, $x(t)$, at some earlier time t plus the velocity times the length of the time interval h. Or, phrased slightly differently,

$$\text{new } x = \text{old } x + \text{velocity} * \text{time interval} \tag{11}$$

Another equally important equation for the subsequent work in this book is obtained from the definition of acceleration. You may recall that the instantaneous acceleration is

$$a(t) = v'(t) = \lim_{h \to 0} \frac{v(t+h) - v(t)}{h} \tag{12}$$

which we will rearrange to give another valuable approximation if h is small enough, namely

$$v(t+h) \simeq v(t) + a(t)h, \tag{13}$$

which says simply: The velocity at a later time is equal to the velocity at an earlier time plus acceleration multiplied by time. Once again, this is an exact equality only if the $v(t)$ versus t curve is a straight line–that is, if a is a constant. We may rephrase the approximation as an equality this way:

$$\text{new } v = \text{old } v + \text{acceleration} * \text{time interval} \tag{14}$$

That is pretty elementary physics, which turns out to be quite powerful physics. Later we will deal with the question of what value of h is small enough.

2.3 About Derivatives and Numerical Integration

> If you are familiar with another programming language and have worked with *subscripted variables*, the notation we have just introduced, **x[n]** and **v[n]**, is Mathematica's version of subscripted variables.

We also need to choose h. Try

In[25]:= **h = 0.5**

Now we can ask Mathematica for the position and velocity at any index n. For example, find **x[10]** and **v[5]**.

In[26]:= **x[10]**
v[5]

The following expressions, which are not to be executed, describe what is happening when Mathematica calculates **v[50]** and **x[50]**, for example. Mathematica computes, in order

$$v[1] = v[0] + a\,h \text{ and } x[1] = x[0] + v[1]\,h$$
$$v[2] = v[1] + a\,h \text{ and } x[2] = x[1] + v[2]\,h \tag{20}$$

and so forth until

$$v[50] = v[49] + a\,h \text{ and } x[50] = x[49] + v[50]\,h \tag{21}$$

In the case of the policewoman in Problem 7 on page 22, the acceleration is constant, but in general it will vary. Then the preceding equations become

$$v[1] = v[0] + a[0]\,h \text{ and } x[1] = x[0] + v[1]\,h$$
$$v[2] = v[1] + a[1]\,h \text{ and } x[2] = x[1] + v[2]\,h \tag{22}$$

and so forth. However, Mathematica does all this for you when you ask for a particular v or x; therein lies some of the beauty and power of a computer with Mathematica on it.

Of course, individual values are of little interest. For the most part we wish to have a more global picture of the velocity. To do this we need to introduce two new Mathematica commands, **Table** and **ListPlot**. We could introduce them simply as required to make the graph we want for this problem. However, it might be more helpful to illustrate with some simpler problems, which we do in the problems below.

Problem 12. We begin with **ListPlot**. Study the output of this instruction.

In[28]:= **ListPlot[{1, 3, 5, 7, 9, 11}];**

What does it graph? Write it in your own words.

Problem 13. Next, try this command:

In[29]:= **ListPlot[{{1, 1}, {2, 4}, {3, 9}, {4, 16}}];**

Describe the difference in the input and output of these last two commands.

Problem 14. Now let's experiment with **Table** to get a better feeling for it. Try this:

In[30]:= **Table[x, {x, 10}]**

At what value of *x* does **Table** start? By what amount does *x* increment for each entry in the table?

Problem 15. Try this command:

In[31]:= **Table[{x, 2 x}, {x, 5, 13, 2}]**

Describe what is different here than in the last **Table** command.

Problem 16. Next, we nest the two commands as follows:

In[32]:= **ListPlot[Table[{x, 2 x}, {x, 5, 10}]];**

Describe what the command did.

Problem 17. Finally, we plot the velocity of the policewoman using **Table** and **ListPlot**.

In[33]:= **ListPlot[Table[{n h, v[n]}, {n, 0, 50}]];**

Carefully explain to yourself each item in this set of two commands so that you understand how Mathematica can graph sets of points that it generates with a **Table** command. For example, what is the significance of the zero in the **{n, 0, 50}** part of the instruction?

We can make this graph much more valuable if we label the axes. We do this as follows:

In[34]:= **ListPlot[Table[{n h, v[n]}, {n, 0, 50}], AxesLabel -> {"t", "v"}];**

> The **AxesLabel -> ....** part of the **Plot** instruction is called an *option*. To find all of the options of a particlar command, type and enter **??ListPlot**, for example. Note that options are always in the form of *rules*; that is, they use the arrow **->** notation as all rules in Mathematica do.

Problem 18. Using the model of **Table** nested in **ListPlot** just provided, plot the position of the policewoman (Problem 7, page 22) as a function of time. It will be easier if you use the cut and paste tools of most editing systems to copy the instruction above, paste it here, and then modify it to graph **x[n]** rather than **v[n]**. You might want to change an axis label as well. Your graph should look like this.

2.3 About Derivatives and Numerical Integration

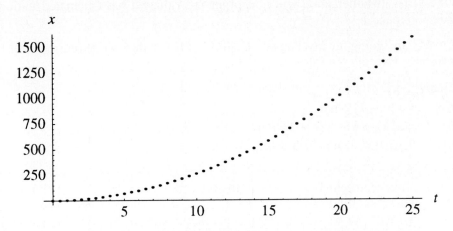

2.3.2 Increasing the Speed of the Calculations

We can achieve a significant increase in the speed with which we can do these calculations with a simple but subtle change in our instructions. This change is neither physics nor mathematics, it is an attribute of Mathematica. Here are the old instructions preceded by a **Clear** instruction.

In[35]:= **Clear[v, x]**
v[n_] := v[n − 1] + a h
x[n_] := x[n − 1] + v[n] h

> Put the previous three commands in one cell, then execute them with a single stroke of the enter key. Cell brackets are shown on the right-hand side of the notebook on the video monitor. In what follows, when you see instructions grouped together, put them in one cell.

Here are the new instructions:

In[38]:= **Clear[v, x]**
v[n_] := v[n] = v[n − 1] + a h
x[n_] := x[n] = x[n − 1] + v[n] h

The result of this change is that Mathematica now remembers each **v[n]** and **x[n]** as it calculates them. For example, with the old instructions, if you wanted **v[15]**, the computer would have to calculate **v[0]** to **v[14]** even if it had calculated them earlier. With the new instructions, if you want **v[15]** and you had, say, calculated **v[10]** earlier, then the computer only needs to calculate **v[11]** through **v[14]** before finding **v[15]**. This saves an enormous amount of time especially when finding the positions, **x[n]**, because those calculations also make use of the **v[n]**. (Mathematica will not forget the values it has calculated until you execute a **Clear** command.)

Problem 19. We are going to investigate the time savings that we promised with the preceding equations. Execute the following old instructions

In[41]:= **Clear[v, x]**
  v[n_] := v[n − 1] + a h
  x[n_] := x[n − 1] + v[n − 1] h
  v[0] = 0; x[0] = 0; a = 5; h = 0.5;

Put the preceding four commands in one cell, then execute them with a single stroke of the enter key. Cell brackets are shown on the right-hand side of the notebook on the video monitor. In what follows, when you see instructions grouped together, put them in one cell.

Next, we do the **ListPlot** nested in a **Timing** instruction, which will tell us how much time it took to execute the instruction. Try it.

In[45]:= **Timing[ListPlot[Table[{n h, x[n]}, {n, 0, 50}], AxesLabel −> {"t", "x"}];]**

Now try the new instructions with the same plot.

In[46]:= **Clear[x, v]**
  v[n_] := v[n] = v[n − 1] + a h
  x[n_] := x[n] = x[n − 1] + v[n − 1] h
  v[0] = 0; x[0] = 0; a = 5; h = 0.5;

In[50]:= **Timing[ListPlot[Table[{n h, x[n]}, {n, 0, 50}], AxesLabel −> {"t", "x"}];]**

We found that with the old instructions it took about 20 times longer to do the plot. Your machine may give quite different results.

2.3.3 More About *h* (Optional)

Problem 20. In this exercise we hope to learn something about the choice of h. Begin with a

In[51]:= **Clear["@"]**

We know that for the policewoman (Problem 7, page 22) the exact value of x is given by $x(t) = \frac{1}{2} a t^2$. Let's compare the exact value with the accurate value. Let

In[52]:= $s[t\_] := \frac{1}{2} a t^2$

$a = 5$

and call a graph of this function **graph1**, like this:

In[54]:= **graph1 = Plot[s[t], {t, 0, 25}];**

Copy the approximation instructions, and make a graph with **ListPlot**, as follows:

In[55]:= **Clear[x, v, h]**
v[n_] := v[n] = v[n − 1] + a h
x[n_] := x[n] = x[n − 1] + v[n] h
v[0] = 0; x[0] = 0; h = 0.5;
graph2 = ListPlot[Table[{n h, x[n]}, {n, 0, 50}], AxesLabel −> {"t", "x"},
   PlotStyle −> {PointSize[.015]}];

Next, we superimpose both graphs.

In[59]:= **Show[graph2, graph1];**

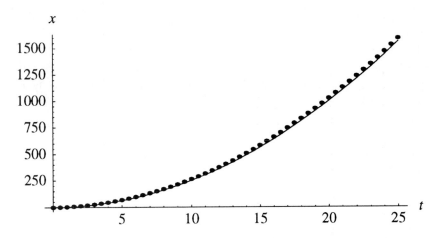

You will see that the approximation for the position of the car is very close to the true position.

Repeat this for some larger values of h. You only need to change h, repeat the approximation calculations, and repeat the **ListPlot** and **Show** instructions. Keep the final value of n in the **Table** instruction equal to $25/h$ so the **Plot** and **ListPlot** graphs have the same time domain. At what value of h does the approximation begin to seriously fail? Clearly, this is a matter of some judgment. Finally, be aware that the results for constant acceleration will, in general, allow a much larger value for h than in those situations where the acceleration varies.

2.4 Quick Summary

Summaries generally come at the end of a chapter, but we wish to repeat and reemphasize what is going on in this chapter for the obvious reason that it is important for you to understand these techniques as well as their motivation.

A great many complex dynamics and kinematics problems, even those that involving accelerations that vary with velocity or position and that cannot be solved by analytic integration, can be solved with these two simple equations:

In[60]:= **v[n_] := v[n] = v[n − 1] + a h**
  x[n_] := x[n] = x[n − 1] + v[n] h

together with initial conditions, such as

In[62]:= **v[0] = 0**
  x[0] = 0

Of course, the acceleration a comes from Newton's second law of motion, $a = F/m$. Problems in thermodynamics, waves, electricity, and magnetism can also be solved with this technique. The two equations are based on an approximation of the derivative (which we repeat just for the velocity)

$$v'(t) \simeq \frac{v(t+h) - v(t)}{h} \tag{23}$$

which may be rewritten as:

$$v(t+h) \simeq v(t) + v'(t)h \tag{24}$$

Note that $v'(t) = a(t)$, and a comes from Newton's second law. The quantity h must be chosen to be small enough to make the calculations valid. The calculations don't make much sense without the ability to plot the results with a **Table** of values and a **ListPlot** command, such as these for the position.

In[64]:= **ListPlot[Table[{n h, x[n]}, {n, 0, 50}], AxesLabel −> {"t", "x"}];**

This particular method of solving differential equations is a modified method of Euler, which we will call Euler-Cromer. Because the output of the process is a table of numbers or a graph, the process is generally called finding a *numerical solution* as opposed to an *analytical solution*, in which case the output is a formula or an equation.

■ 2.5 About Our Programming Methods

The computer programming technique we are using in Mathematica is called *recursion*. New values of a quantity are defined in terms of previous values, which in turn must ultimately be defined in terms of some inital value or values. Perhaps the most famous example of recursion is the calculation of terms in the Fibonacci sequence: 1, 1, 2, 3, 5, 8, . .., in which each term except the first two are defined in terms of the sum of the two previous terms. In Mathematica we can express this recursively as follows:

In[65]:= **Clear[f]**
   f[1] = 1; f[2] = 1;
   f[n_] := f[n] = f[n − 2] + f[n − 1]

Then we try calculating a few terms.

In[67]:= **f[4]**

In[68]:= **f[6]**

> A difficulty with recursion as a technique in Mathematica occurs when one or more parts of the definition are absent or incorrect (for example, an initial value definition is absent, or *h* is undefined); then you get either an error message or the recursion goes on and on, seemingly forever. If that appears to be the case with your calculations, learn how to abort a Mathematica calculation for your particular computer. Otherwise, recursion is a marvelously efficient way to handle the numerical integration calculations we wish to perform. Another bit of advice when using recursion is to **Clear** often, usually *every time* you execute the recursive commands.

Problem 21. You might try writing a recursive function that calculates **n!**. If you have forgotten, $n! = n(n − 1)(n − 2) \ldots 1$. For example, $3! = 6$, $4! = 24$, and $5! = 120$. Note that by definition $0! = 1$ and $1! = 1$.

One way to check our recursion method of finding $v(t)$ and $x(t)$ given an acceleration is to take a case that can be integrated analytically and compare it with the numerical approach. Suppose the force on a 1-kg mass varies with time in such a way that $F = t^2$. Then $a = F/m = t^2$. Suppose further that at $t = 0$, $v = 0$ and $x = 0$. Then the kinematics integrals are

In[69]:= $v = \int_0^t t^2 \, dt + 0$

$x = \int_0^t v \, dt + 0$

(Your experience with calculus should enable you to integrate these equations without Mathematica.) Therefore, the analytic solutions are $v = t^3/3$ and $x = t^4/12$. Next we try our numerical approach, and then we will compare the methods with graphs. Recall that $t = nh$. We write

In[71]:= **Clear["@"]**
a[t_] := t^2
v[n_] := v[n] = v[n − 1] + a[(n − 1) h] h
x[n_] := x[n] = x[n − 1] + h v[n]
x[0] = 0; v[0] = 0; h = 1;

To plot our numerical solution, we write

In[76]:= **numericalgraph =**
  ListPlot[Table[{n h, x[n]}, {n, 0, 100, 2}], Prolog −> {PointSize[0.015]},
  AxesLabel −> {"t", "x"}, PlotRange −> All];

where we have more or less arbitrarily chosen a final n of 100, or a time $t = nh = 100$ s. To graph our analytic solution, we use a **Plot** instruction. Thus

In[77]:= **analyticgraph = Plot$\left[\dfrac{t^4}{12}, \{t, 0, 100\}, \text{PlotRange} \to \text{All}\right]$;**

We superimpose the two to illustrate that our solution to $t = 100$ is quite good.

In[78]:= **Show[numericalgraph, analyticgraph];**

Next, we try $h = 20$ and graph the two solutions over the same interval. All of the instructions are grouped together, and we have kept only the superposition graph. Clearly, $h = 20$ begins to cause significant error in an interval of 100 s. Errors also accumulate and, therefore, they increase with the length of the integration interval.

In[79]:= **Clear[v, x, h]**
a[t_] := t^2
v[n_] := v[n] = v[n − 1] + a[(n − 1) h] h
x[n_] := x[n] = x[n − 1] + h v[n]
x[0] = 0; v[0] = 0; h = 20;

```
numgraph =
    ListPlot[Table[{n h, x[n]}, {n, 0, 5, 1}], Prolog -> {PointSize[0.015]},
        AxesLabel -> {"t", "x"}, PlotRange -> All];
Show[numgraph, analyticgraph];
```

Problem 22. Repeat this process for $v(t)$ rather than $x(t)$. Explain why h needs to be smaller to get good results for x than it does for v.

■ 2.6 A Final Illustration of This Technique

Let's try our method one more time in this chapter before we move to the next chapter where we will apply the method to falling raindrops and other objects. For our final illustration in this chapter we will assume the acceleration of our particle depends on position, as follows:

$$a = -x \qquad (25)$$

This acceleration could follow from a force law where $F = -kx$ with $k = 1$ N/m with F acting on a mass of 1 kg. Newton's second law becomes $-kx = ma$, and if $k = m = 1$, then $a = -x$. Many springs and Bungee cords produce forces similar to this, at least over part of their stretching domain.

Depending on the initial conditions, the solution to this problem is known from mathematics to be sine and cosine functions. We will see if our equations also give sine and cosine solutions. Here are our equations modified for this situation:

```
In[84]:=  Clear[x, v, a]
          a[x_] := -x
          v[n_] := v[n] = v[n - 1] + a[x[n - 1]] h
          x[n_] := x[n] = x[n - 1] + v[n] h
          h = 0.1;
```

where we have chosen $h = 0.1$. Let's take for initial conditions

```
In[89]:=  v[0] = 1; x[0] = 0;
```

which means we are starting the particle at $x = 0$ with a velocity kick of some kind.

Here are some provisions to plot our data as follows:

```
In[90]:=  graph1 = ListPlot[Table[{n h, v[n]}, {n, 0, 63}]];
          graph2 = ListPlot[Table[{n h, x[n]}, {n, 0, 63}]];
```

Kowa Bungee!

Problem 23. Can you figure out a way to superimpose sine and cosine curves on these graphs to see how well the points fit the true curves? It is likely you will need the **Plot** and **Show** instructions; also remember $t = nh$.

Problem 24. If you made a graph of **v[n]** versus **x[n]**, what would you expect to get? Make a prediction. Now use **ListPlot** and **Table** to make such a graph, called a phase-space plot, to verify your prediction. Plot from $n = 0$ to $n = 50$. On your graph identify the first point. How do you know that this is the first point as opposed to the last point?

Mostly Mathematica

1. If

$$f(x) = -x^2 - 2x + \pi, \qquad (26)$$

Create a function for f in Mathematica and graph the function.

2. What are the zeros of f? Where is f a maximum? At what value of x is $f(x) = 1$? What is $f'(x)$? What is $f''(x)$? Is $f''(x)$ ever zero?

3. Given that $0! = 1$, $1! = 1$, and $n! = n(n-1)(n-2) \ldots 1$, write a recursion relationship that gives you n factorial ($n!$).

Explorations

1. The equations we used in the preceding section, namely

```
In[91]:= v[n_] := v[n] = v[n - 1] - x[n - 1] h
         x[n_] := x[n] = x[n - 1] + v[n] h
```

lack a certain element of symmetry. What happens to the solution to this problem in Section 2.6 if you incorporate the symmetry as follows, changing **v[n]** to **v[n − 1]** in the second equation?

```
In[93]:= v[n_] := v[n] = v[n - 1] - x[n - 1] h
         x[n_] := x[n] = x[n - 1] + v[n - 1] h
```

You may also wish to explore the article by Cromer (1981).

2. Study the behavior of a Duffing oscillator, which has a force law $F = x - x^3$. Assume the force is acting on a 1-Kg mass. For initial conditions you might try $x(0) = 1$ and $v(0) = -1$ as well as some other values. The phase plot, $v(t)$ versus $x(t)$, will be the most valuable plot, but also look at $v(t)$ versus t and $x(t)$ versus t. You might begin by graphing the force as a function of x. For what values of x is the force toward the origin? Away from the

2.6 A Final Illustration of This Technique

origin? What will happen if the initial conditions are $x = \pm 1$ and $v = 0$? What will happen if the initial conditions are $x = 0$ and $v = 0$?

3. Study the motion of a 1-Kg mass under the influence of a force

$$F(x) = \left(-\frac{1}{x^2} + \frac{1}{x^3}\right) \qquad (27)$$

The form of this force is similar *in form* to that between atoms in certain molecules. Begin by graphing the force with the **Plot** instruction. You will need to find some initial conditions that work and vary them to see the various possibilities of this force law.

4. The technique we have developed in this chapter is applicable to many kinds of problems, as long as the derivative of a function is known. For example, the rate of change of temperature of some warm liquid in a container in a cooler environment is proportional to the temperature difference between the liquid and the environment. Thus

$$\frac{dT}{dt} = k(T - T_e) \qquad (28)$$

where T is the temperature of the liquid and T_e is the constant environmental temperature. To find a numerical solution we use the derivative approximation once again and make repeated use of the approximation

$$T(t + h) \simeq T(t) + \frac{dT}{dt} h \qquad (29)$$

In Mathematica, we write

In[95]:= **Clear["Global`*"]**
Tprime[t_] := −k (t − t$_e$)
T[n_] := T[n] = T[n − 1] + Tprime[T[n − 1]] h

for our recursion relationship. Note that we need some initial value for T, a value for k and the outside temperature, T_e. To experiment with this problem you might take $T(0) = 80\ °C$, $T_e = 20\ °C$, $h = 1$ s, $k = 0.01$. You might also set up a real experiment in which you measure the temperature of some hot water in a cup as a function of time and see if the theory agrees with the experiment. For further details, see Gould and Tobochnik (1988).

5. In this exploration we give only a rough outline of what you might do, leaving all of the details to you. In a number of physical and biological systems the derivative of the number of things (atoms, rabbits, money in the bank, etc.) is proportional to the number of things. Thus

$$\frac{dN}{dt} = kN \qquad (30)$$

where N will increase if $k > 0$ (population or money in the bank) but decrease if $k < 0$ (radioactive decay). Find some numerical solutions to this problem for k both positive and negative. Find this equation in your calculus text and compare your numerical solution to the exact solution.

References

Cromer, A., Stable solutions using the Euler approximation, *Am. J. Phys.* 49 (1981): 455.

Epstein, L.C., *Thinking Physics*, San Francisco: Insight Press, 1983, p.165.

Gould H. and J. Tobochnik, *Computer Simulation Methods*, Reading, Mass.: Addison-Wesley, 1988, p.12.

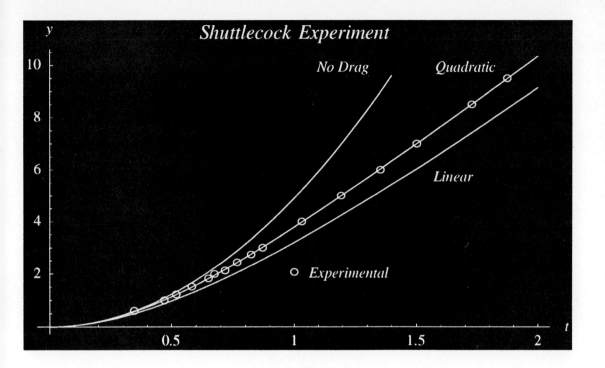

CHAPTER 3 Raindrops, Pebbles, and Shuttlecocks: Objects Falling in Air

■ 3.1 Introduction

Think about something falling in air, a pebble, raindrop, or sky diver. It seems clear that the faster something travels through the air, the greater the resistive force (drag) of the air. Thus we might write

$$F \propto v^r \qquad (1)$$

for the magnitude of the force where r is a real number. Morever, the force of the air is always in a direction opposite to the direction of the velocity. In some cases the magnitude of the drag force can be written

$$F = \tfrac{1}{2} C \rho_a v^2 \qquad (2)$$

where C is called the drag coefficient, ρ_a is the density of air, A is the cross section of the object perpendicular to v, and v is the magnitude of the velocity. This may seem mysterious, but it turns out that

$$\tfrac{1}{2}\rho_a A v^2 \qquad (3)$$

is simply the force required to accelerate a cylindrical air mass with cross section A from zero to speed v in a distance equal to the length of the cylinder. We know this is only an approximation to the drag force, so we include another factor, C, the drag coefficient, which depends mainly on the geometric properties of the object.

Draw a free body diagram for a falling mass. Then, taking downward as positive, Newton's second law becomes

$$mg - \tfrac{1}{2}C\rho_a v^2 = ma, \qquad (4)$$

and solving for the acceleration gives

$$a = g - \frac{C\rho_a A}{2m}v^2 = g - kv^2 \qquad (5)$$

with

$$k = \frac{C\rho_a A}{2m} \qquad (6)$$

We will enter this information into our numerical integration scheme (introduced in Chapter 2) to find a numerical solution, hoping to learn something about objects falling in a resistive medium. We will take as our initial condition $v = 0$, as if we were dropping the object rather than throwing it downward.

Problem 1. Certainly you have some idea of how the velocity will change as an object falls through the air. At least make an attempt at a prediction by sketching a graph of velocity (downward) as a function of time. Remember your initial condition is $v = 0$, so your graph should go through the origin, and remember that the drag increases the faster the object drops. You may use this coordinate system for your prediction.

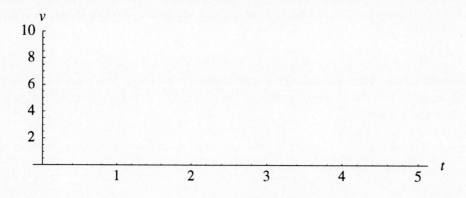

3.2 On Raindrops and Pebbles

> In this chapter and in all notebooks and chapters to follow, you will be expected to enter and execute the Mathematica commands in the sequence in which they are given. Failure to do so may produce error messages and meaningless results.

In this section we will create some mathematical models of objects falling through air. Simply to get a preliminary idea of what our model predicts for how the velocity changes with time, we choose the constant k to be 0.1. Here are the equations to solve this problem.

> Put the following commands in one cell, then execute them with a single stroke of the enter key. Cell brackets are shown on the right-hand side of the notebook on the video monitor. In what follows, when you see instructions grouped together like this, put them in one cell.

```
Clear["@"]
a[v_] := g - k v^2
v[n_] := v[n] = v[n - 1] + a[v[n - 1]] h
h = 0.1; g = 9.8; k = 0.1; v[0] = 0;
ListPlot[Table[{n h, v[n]}, {n, 0, 50}], AxesLabel -> {"t", "v"}];
```

Try it and see how it compares with your predicted curve. Then proceed to the next group of problems.

> If you wish to repeat these instructions after varying one or more of the parameters, be sure to first perform a **Clear** command; otherwise Mathematica will simply remember the old values.

Problem 2. How does the velocity of a falling object change if the air resistance is zero? Without using Mathematica, sketch what you think is a graph of v versus t for this case.

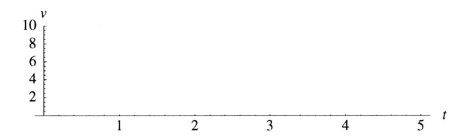

Write the equation of v as a function of t for an object falling in a vacuum, then graph the function using Mathematica's **Plot** command. How did you do with your prediction?

Problem 3. What is the most significant difference between the v versus t graphs for free-fall in a vacuum and falling in air?

Problem 4. Define what you think terminal velocity means.

Problem 5. What is the value of the terminal velocity for the object in our model?

Problem 6. How does the terminal velocity change with the constant k? In other words, what if $k = 2$? What if $k = 0.5$? You are allowed to repeat the calculations with a different values of k, but first make some predictions about what you expect will happen.

Problem 7. How do you think the acceleration of the falling object varies with time? Make a prediction by drawing a graph of a versus t for our falling object.

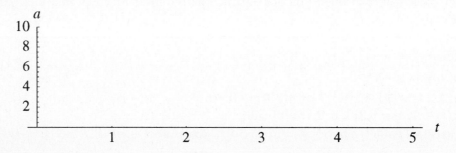

Problem 8. Here is an instruction that will make a plot of the acceleration versus time for our Mathematica model. Execute it and compare the results of the model with your prediction. If you cleared some variables, you may have to rerun the model with our trial value for k, namely $k = 0.1$

$$\text{ListPlot[Table[\{n h, (g - k v[n]}^2)\}, \{n, 0, 50\}], AxesLabel -> \{"t", "a(t)"\}];$$

What is the initial value of the acceleration? What is the limiting value of the acceleration, that is, what is a when t becomes very large?

Problem 9. In the system of equations for the Mathematica model of the falling object, add an equation to calculate the position as a function of time, then plot the position as a function of time. (You may need to refer to the previous chapter to see how to find $x(t)$, and be sure to include the initial condition **x[0] = 0**.) What are the main differences between this curve and the curve for an object falling in a vacuum?

Now let's talk about a real object; in particular we will assume we are dealing with a spherical raindrop whose radius is 1 mm. Since

$$m = \rho \tfrac{4}{3} \pi r^3 \qquad (7)$$

3.2 On Raindrops and Pebbles

is the mass of our raindrop and the density of water is 1000 kg/m³, then, in Mathematica

$$r = .001; \rho = 1000; m = 4/3 \, \rho \, \pi \, r^3$$

Also assume that $C = \frac{1}{2}$, the density of air is 1.2 kg/m³, in which case the constant k becomes

$$c = 1/2; \rho_a = 1.2; k = \frac{c \, \rho_a \, \pi \, r^2}{2 \, m}$$

> We have used a lower-case **c** in this equation because the upper-case **C** is a reserved word in Mathematica.

Now we apply our equations to the raindrop.

```
Clear[v]
a[v_] := g - k v^2
v[n_] := v[n] = v[n - 1] + a[v[n - 1]] h
h = 0.05; g = 9.8; v[0] = 0;
ListPlot[Table[{h n, v[n]}, {n, 0, 50}], AxesLabel -> {"t (s)", "v(t) (m/s)"}];
```

Problem 10. What is the terminal velocity of our hypothetical raindrop? Suppose the raindrop has half the radius as the one we modeled, what do you predict will be the effect on its terminal velocity as compared to the 1-mm drop? Model its fall. Serway (1996) lists 9.0 m/s as the terminal velocity of a raindrop with radius 2 mm. What terminal velocity does our model predict?

Problem 11. Does the terminal velocity increase or decrease with the size of the drop? Could you have predicted this from a simple consideration of how the cross-sectional area to mass ratio changes with r? Explain. Do fog droplets fall? Why or why not?

Problem 12. Will a pebble fall faster than a raindrop? How do you know?

Problem 13. When the falling object reaches terminal velocity, the acceleration is zero. Use this to show that the magnitude of the terminal velocity is given by

$$v_T = \sqrt{\frac{2 \, mg}{C \rho_a \, A}} \qquad (8)$$

For a more challenging exercise in analysis, show that for *spherical* raindrops of the same density the terminal velocity increases with the square root of the radius. Hint: substitute for m and A as functions of r in Equation 8.

Problem 14. Given the results in Problem 13, show that Newton's second law for the falling object becomes

$$a = g\left(1 - \left(\frac{v}{v_T}\right)^2\right) \qquad (9)$$

Problem 15. Which falls faster, a hailstone or a water drop of the same size? How do you know?

Problem 16. Compare the speeds of raindrops that fall from a height of 100 m in a vacuum with those that fall in air. Choose raindrops with $r = 1$ mm and $r = 2$ mm. Convince yourself that both drops in air will reach terminal velocity before falling 100 m.

■ 3.3 Throwing Rocks

Problem 17. We are going to model the situation where a rock is thrown straight up into the air. If you were to compare the heights achieved by a rock thrown straight upward in a vacuum and one thrown in air, which do you predict would go higher, assuming they both start with the same initial velocity? Which do you think will stay in the air longer? Explain your thinking.

Problem 18. Draw two graphs on the following coordinate system, the first should show how the velocity of a rock thrown upward changes with time if there is no air resistance, and the second should show the nature of this curve with air resistance. These are predictions; there is no penalty for being wrong. Assume the rock has an initial upward velocity of 25 m/s.

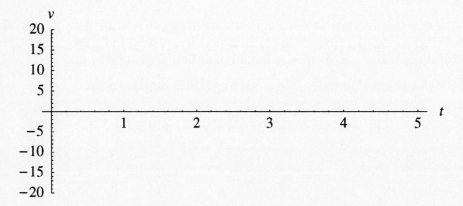

We must make some adjustments in the equations of the previous section because on the way up the drag force on the rock is down, while on the way down the drag force is up. Our previous equations did not take that into account. Here is something that works for this situation:

3.3 Throwing Rocks

$$a = -g - kv|v| \qquad (10)$$

where k is the constant defined earlier.

Assume a spherical rock of radius $r = 0.01$ m, an air density of 1.2 kg/m^3, and the density of rock equal to 2000 kg/m^3. Then, taking plus as upward, the Mathematica commands become

```
Clear["@"]
c = 1/2; ρ = 2000; r = 0.01; k = c 1.2 π r² / (2 ρ 4/3 π r³);
a[v_] := -g - k v Abs[v]
v[n_] := v[n] = v[n - 1] + a[v[n - 1]] h
y[n_] := y[n] = y[n - 1] + v[n] h
v[0] = 25; y[0] = 0; h = .1; g = 9.8;
g1 = ListPlot[Table[{n h, y[n]}, {n, 0, 42}],
      AxesLabel -> {"t (s)", "y (m)"}];
```

Problem 19. Now you can answer the questions at the beginning of the section: Which goes higher and which stays in the air longer, a rock thown upward in air or one thrown upward in a vacuum? To answer this convincingly, you may have to recall or look up the formula for the position of an object thrown upward in a vacuum. What is this formula? Now superimpose a graph of this latter equation made using the **Plot** command on the graph constructed by our Mathematica model. You will need a **Show[g1, g2]** command, where **g1** is the graph of the Mathematica model and **g2** is the graph you just constructed.

Problem 20. Make a graph of the velocity of the rock versus time and compare this with your previous prediction. If you went wrong, where did you go wrong?

Problem 21. You can try to predict how the acceleration varies with time for our rock, but it is not easy. What do you predict is the acceleration of the rock when it is at its highest point? Make a prediction before you execute the next command, which will plot the acceleration. Interpret the various features of the graph. Is the acceleration zero at the top of the flight of the rock? Will a rock thrown upward ever reach terminal velocity on its way down?

```
ListPlot[Table[{n h, a[v[n]]}, {n, 0, 42}], AxesLabel -> {"t", "a(t)"}];
```

Problem 22. Suppose you have two rocks of the same density but different sizes and you give each the same initial velocity upward. Which will go the highest? Suppose you have two rocks of the same size but different densities. Assuming you have the strength to give each the same initial velocity, which rock can you throw the highest?

■ 3.4 Testing the Model

It would be nice to test our ideas and models with some real experimental data, and in fact we can do that. We have some distance-time data on a falling shuttlecock from some fine electronic measurements by Peastrel, Lynch, and Armenti (1980). We have assembled them in a list called **data**.

data = {{0.347, 0.61}, {0.47, 1.}, {0.519, 1.22}, {0.582, 1.52},
{0.65, 1.83}, {0.674, 2.}, {0.717, 2.13}, {0.766, 2.44},
{0.823, 2.74}, {0.87, 3.}, {1.031, 4.}, {1.193, 5.},
{1.354, 6.}, {1.501, 7.}, {1.726, 8.5}, {1.873, 9.5}}

This list, with time in seconds in the first column and distance dropped in meters in the second column, is more easily studied if we put it in the form of a table with the **TableForm** instruction, as follows:

TableForm[data]

This experimental data may be graphed with the command

ListPlot[data, AxesLabel –> {"t", "y "}];

The shuttlecock was dropped and the distance was measured from the drop point, so the initial conditions are $v = y = 0$ at $t = 0$. Recall from Problem 14 that the acceleration may be described in terms of a single parameter, the terminal velocity v_T, with the expression

$$a = g(1 - \left(\frac{v}{v_T}\right)^2). \tag{11}$$

What we would like to know is if there is a terminal velocity that gives a good fit between the model and the experimental data. To answer that we can graph the data and the model predictions on the same coordinate system and make changes in the terminal velocity until we see if a good fit is possible. Since we will repeatedly calculate new model predictions for different values of the terminal velocity, we will group all of our instructions in a single **Module** with the terminal velocity as a parameter. The number of points, **np**, and the time interval **h** will also be input parameters.

3.4 Testing the Model

```
modelcheck[vterm_, np_, h_] := Module[{n},
  Clear[y, v];
  a[v_] := g (1 - v Abs[v] / vterm^2);
  v[n_] := v[n] = v[n - 1] + a[v[n - 1]] h;
  y[n_] := y[n] = y[n - 1] + v[n] h;
  v[0] = 0; y[0] = 0; g = 9.8;
  model = Table[{n h, y[n]}, {n, 0, np, 10}];
  ListPlot[Union[model, data], AxesLabel -> {"t", "y"}];]
```

Here is how you find and graph a set of points from the model and from the experimental data. We have chosen an h of 0.01 s, a terminal velocity of 15 m/s, and we plot 190 points: 190 points gives a time of $190h = 1.9$ s, which corresponds approximately in time to the last experimental data point. Execute this command:

modelcheck[15, 190, .01]

Problem 23. Repeat runs of **modelcheck** until you find a terminal velocity that gives good agreement between the experimental data and the model. In your opinion is the agreement good?

Problem 24. Here is some data from Greenwood, Hanna, and Milton (1986) on a falling Styrofoam ball with $r = 0.0254$ m and $m = 0.00254$ kg. What terminal velocity do you get from a fit between the model and the experimental data? Also find the drag coefficient, C.

data = {{0.0, 0.075}, {3/30, 0.260}, {6/30, 0.525}, {9/30, 0.870},
 {12/30, 1.27}, {15/30, 1.73}, {18/30, 2.23}, {21/30, 2.77}, {24/30, 3.35}}

Mostly Mathematica

1. Suppose the acceleration of an object is

$$a(t) = -t \qquad (12)$$

Using the numerical integration formulas we have developed, find the object's subsequent position and speed if $x(1) = 0$ and $v(0) = 5$ for $t = 0$ to 4. Use $h = 0.1$. Also find the subsequent position and speed analytically using Mathematica's integration commands. What is the percentage difference between the numerical and analytic values for the final position?

2. Refer to the previous problem and find the distance traveled by the object from $t = 0$ to $t = 4$ using both analytical and numerical integration methods. (Note: The distance traveled is not the difference $x(4) - x(0)$. Graphs of $x(t)$ may help.)

3. Find, both analytically and numerically when the object has a velocity of zero. Note that the numerical value might not be exactly zero. At what instant between $t = 0$ and $t = 4$ is the position a maximum?

Explorations

1. We have assumed that the force of air resistance varies directly with the square of the velocity. Explore how the velocity and position of a falling object vary with time when the drag force varies with both smaller and larger powers of v. Sometimes the trick in the equations is to keep the acceleration due to the air resistance opposite to the velocity, but as long as you deal simply with a falling object, the force of air resistance will be upward. Can you get good fits between the experimental data and the model if the air resistance varies directly with the velocity, a situation that occurs at low speeds?

2. What are the terminal velocities of bugs and small mammals? Would a bug dropped from the top of a building be killed? What are some masses and cross-sectional areas?

3. If your department has a sonic motion detector, make some distance versus time measurements on falling coffee filters and try to find a model that fits the data. Does a linear model, $F_a \propto v$ or a quadratic model, $F_a \propto v^2$, fit the data better? You can also get data from Videopoint if your department has this software package.

4. In a subsequent chapter we will show that both the linear and quadratic models have analytic solutions. A model having both a linear drag term and a quadratic drag term does not have an analytic solution. If the drag force is quadratic in v, then the solution for both $v(t)$ and $y(t)$ of a falling object with $y = v = 0$ at $t = 0$ is

$$v(t) = v_T \tanh\left(\frac{gt}{v_T}\right) \text{ and } y(t) = \frac{v_T^2}{g} \ln\left(\cosh\left(\frac{gt}{v_T}\right)\right) \quad (13)$$

Yes, those are the hyperbolic functions, the functions you never thought you would see, and v_T is the terminal velocity.

If the drag force is linear in v, then the solution is

$$v(t) = v_T\left(1 - e^{-\frac{gt}{v_T}}\right) \text{ and } y(t) = \frac{v_T^2}{g}\left(\frac{gt}{v_T} - 1 + e^{-\frac{gt}{v_T}}\right) \quad (14)$$

Use Mathematica's **Plot** command and the graphs obtained from our models to compare the model predictions with the theoretical results, especially those for the shuttlecock. Refer to the graph at the chapter heading to see our comparison.

References

Greenwood, M. S., C. Hanna, and J. Milton, Air resistance acting on a sphere, *The Physics Teacher* **24** (1986): 153.

Peastrel, M., R. Lynch, A. Armenti, Jr. Terminal velocity of a shuttlecock in vertical fall, *Am. J. Phys.* 48 (1980): 511.

Serway, R. A., *Physics for Scientists and Engineers,* Philadelphia: Saunders College, 1996, p. 157.

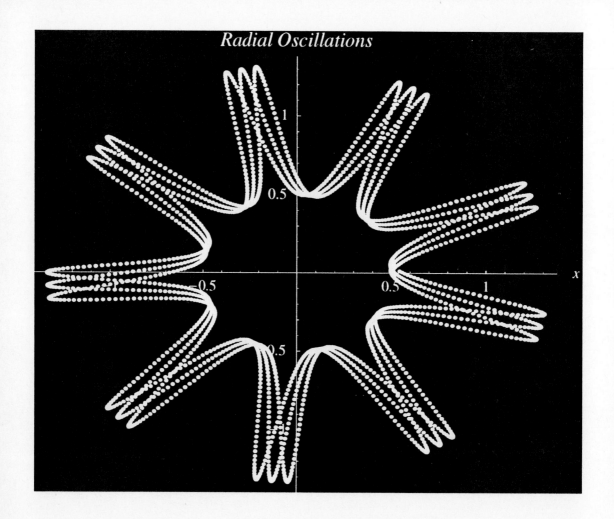

CHAPTER 4 Vectors, Baseballs, Planets, and Moon Shots

■ 4.1 Introduction

In this chapter we advance to two-dimensional physics through the use of vectors. This allows us to consider one of the most important forms of motion in the history of physics, the motion of the planets. With the use of vectors our numerical methods allow us to attack much more interesting problems. We will introduce Mathematica's unique vector notation and the dot product, and we will show how to identify and isolate the components of a vector. Then we will apply the Euler-Cromer method and a variation of it to two-dimensional Newton's second law problems. Finally, we will come up against the limits of this method,– namely precision, computer time, and memory requirements.

■ 4.2 Simple Vector Operations in Two Dimensions

You have studied vectors and vector operations such as addition, subtraction, and perhaps the dot product. For example, if vector **A** had components A_x and A_y and vector **B** has components B_x and B_y, then **A** + **B** has components A_x+B_x and A_y+B_y and **A** − **B** has components A_x-B_x and A_y-B_y. In **i**, **j** notation, if **A** = A_x**i** + A_y**j** and **B** = B_x**i** + B_y**j**, then **A** + **B** = (A_x+B_x)**i** + (A_y+B_y)**j**. Mathematica uses a slightly different notation, probably somewhat similar to what you used in your calculus course. In calculus, we typically write **A** = (A_x, A_y) or **A** = $<A_x, A_y>$, while in Mathematica we write **A** = $\{A_x, A_y\}$; set brackets are used for sets, vectors, and matricies. Some examples will clarify this for vectors. Let

In[1]:= **a = {1, −2};**
b = {0, 3};

Then we can find the sum and difference as follows:

In[2]:= **a + b**
a − b

How can we extract the components of **a**? Mathematica has unusual notation for this. To find the *x*-component of **a**, we write

In[4]:= **a[[1]]**

while the *y*-component is

In[5]:= **a[[2]]**

Therefore, if we want the magnitude of **a**, we ask for

In[6]:= $\sqrt{\text{a[[1]]}^2 + \text{a[[2]]}^2}$

If **r** = {1, 5} and **s** = {2, −7}, find the magnitude of **r** + **s**. We can do this in one step, as follows:

In[7]:= **r = {1, 5};**
s = {2, −7};
$$\sqrt{(r+s)[[1]]^2 + (r+s)[[2]]^2}$$

Problem 1. Find the magnitude of **a**, **b**, **a** − **b** and **b** − **a** where **a** = {3, 4} and **b** = {−5, 2}.

Problem 2. Show with a counterexample that vector subtraction is not commutative.

> Vector operations in three dimensions are essentially the same except we have three components. Thus, for example, **a** = {1, −1, 5} is a vector in 3-space.

■ 4.3 The Dot Product

Let's clear our variables and then discuss the dot (or scalar) product.

In[8]:= **Clear["@"]**

Let

In[9]:= **a = {1, 3};**
b = {2, −5};

Mathematica performs the dot product with this instruction

In[10]:= **a.b**

The magnitude of a vector can be found easily with the dot product:

In[11]:= $\sqrt{\mathbf{a.a}}$

gives the magnitude of **a**. We could easily create a command to find the magnitude of a vector. Here it is, in fact.

In[12]:= **mag[x_] := $\sqrt{\mathbf{x.x}}$**

We use our new command to find the magnitude of **a**.

In[13]:= **mag[a]**

What is the angle between **a** and **b**? You may recall that the dot product can be used to find this angle. Use

In[14]:= $\theta = \text{ArcCos}\left[\dfrac{\mathbf{a.b}}{\sqrt{\mathbf{a.a}}\ \sqrt{\mathbf{b.b}}}\right]$

to get an exact value or

In[15]:= $\theta = \text{ArcCos}\left[\dfrac{\mathbf{a.b}}{\sqrt{\mathbf{a.a}}\ \sqrt{\mathbf{b.b}}}\right]$ // N

to get a decimal approximation (in radians). Let's try this with the familiar unit vectors, **i** and **j**. First we define them in Mathematica notation.

In[16]:= **i** = {1, 0};
       j = {0, 1};

Now we find the angle between them,

In[17]:= $\theta = \text{ArcCos}\left[\dfrac{\mathbf{i.j}}{\sqrt{\mathbf{i.i}}\ \sqrt{\mathbf{j.j}}}\right]$

which should be the angle we expect. Note that this technique always gives an angle θ where $0 \leq \theta \leq \pi$.

Problem 3. Find the angle between **i** and **–i**.

Problem 4. Find the decimal approximation of the angle in radians which the vector **d** = {–3, –4} makes with the –x-axis (–**i**). Can you tell what quadrant an angle is in using the dot product? How?

Problem 5. A projectile has an initial velocity **v** = {5, 3}. What is the magnitude of the initial velocity? What angle does the initial velocity make with the +x-axis?

■ 4.4 Projectile Motion (in a Vacuum)

Here are the projectile motion formulas found in almost any physics textbook, but arranged in Mathematica's vector form.

In[18]:= **Clear["@"]**
    a = {0, –g}
    v = {30, 40 – g t}
    r = {30 t, 50 t – 1/2 g t²}

where we have assumed an initial velocity of {30, 40}. Let's assign a value to *g;*

In[22]:= **g = 9.8;**

and find when the projectile hits the ground. It does this when the *y*-component of **r** is zero.

In[23]:= **Solve[r[[2]] == 0, t]**

The first solution, $t = 0$, is when the projectile starts, so the second value, $t = 10.2041$, must be when $y = 0$ once again. We can make a graph of the trajectory:

In[24]:= **ParametricPlot[r, {t, 0, 10.2}, AxesLabel –> {"x", "y"}];**

> Do'nt worry about the error message associated with this **ParametricPlot** command.

We can also find the height it goes, knowing that the time it takes going up is half of the total time.

In[25]:= **r /. t –> 10.20401 / 2**

We can even find derivatives; the first derivative of **r** should be **v**:

In[26]:= ∂_t **r**

The second derivative gives us **a**:

In[27]:= $\partial_{t,t}$ **r**

where, in this case, we knew the answers before we started. How far does the projectile travel through the air? The answer is

In[28]:= $\int_0^{10.2041} \sqrt{\mathbf{v}\cdot\mathbf{v}}\, d t$

Problem 6. A projectile is launched on the Moon with an initial velocity whose magnitude is 10 m/s at an angle of 30 degrees with respect to the horizontal. The acceleration of gravity on the Moon is about *g*/6. Create an acceleration vector **a** = {0, –9.8/6} and an initial velocity vector **vo** = {10 Cos[30 Degree], 10 Sin[30 Degree]}. Then integrate the acceleration vector to get **v**, as follows:

4.5 A Vector Approach to the Euler-Cromer Method

```
In[29]:= a = {0, -9.8 / 6};
       vo = {10 Cos[30 Degree], 10 Sin[30 Degree]}
```

$$v = \int_0^t a\,dt + vo$$

Now integrate **v** to find **r**, given that **ro** = {0, 0}.

```
In[31]:= r =
```
$$\int_0^t v\,dt + \{0, 0\}$$

Finally, plot the trajectory using **ParametricPlot**.

Problem 7. Using the results of Problem 6, find how high the projectile goes. Also find its range. How long does it stays in the air? How far does it travel?

Problem 8. How much higher does this projectile go on the Moon than it would go on the Earth? By what factor is its range reduced on Earth?

■ 4.5 A Vector Approach to the Euler-Cromer Method

```
In[32]:= Clear["@"]
```

We might profitably review our numerical analysis approach to solving dynamics problems, in this case using vectors for motion in two dimensions. Our starting point is Newton's second law, which, given the forces acting on a particle, gives us the acceleration of the particle. Then we use the definition of acceleration,

$$\mathbf{a}(t) = \lim_{h \to 0} \frac{\mathbf{v}(t+h) - \mathbf{v}(t)}{h} \quad (1)$$

which for sufficiently small h and some rearranging becomes

$$\mathbf{v}(t+h) \simeq \mathbf{v}(t) + h\mathbf{a}(t) \quad (2)$$

In Mathematica this formula is applied in steps as follows:

$$\begin{aligned}\mathbf{v}[1] &= \mathbf{v}[0] + h\mathbf{a}[0]\\ \mathbf{v}[2] &= \mathbf{v}[1] + h\mathbf{a}[1]\end{aligned} \quad (3)$$

until we have found as many values of **v** as we wish. The general formula is

$$\mathbf{v}[n] = \mathbf{v}[n-1] + h\mathbf{a}[n-1] \quad (4)$$

We do a similar thing with the velocity, beginning with the definition of **v**, to find **r**(*t*). You might try supplying the intermediate steps to show how we obtain the recursion formula for **r**, which follows:

$$r[n] = r[n-1] + hv[n] \qquad (5)$$

We call this numerical integration method the Euler-Cromer method. It can profitably used with any problem where the derivative of a function is known; the approximation is always of the form

$$f(x + h) \simeq f(x) + hf'(x). \qquad (6)$$

Before attacking the problem of a projectile with air resistance, we apply the methods we described in the last chapter to the two-dimensional problem of a projectile traveling through a vacuum. In this case the acceleration, expressed as a vector in Mathematica, is

In[33]:= **a[n_] := {0, −g}**

where *g* is the acceleration of gravity. Note that in this case **a** is a constant vector and does not depend on *n*. Our recursion formulas for **r** and **v** look exactly like they did before, but Mathematica will treat them as vectors. Here they are: One for position

In[34]:= **r[n_] := r[n] = r[n − 1] + h v[n]**

and one for velocity

In[35]:= **v[n_] := v[n] = v[n − 1] + h a[n − 1]**

Finally, we need to give initial conditions and give all parameters numerical values, thus

In[36]:= **r[0] = {0, 0}; v[0] = {3, 4}; g = 9.8; h = 0.01**

Here is everything collected into one command, including a command to plot our *x-y* results.

> Put the following instructions in a single cell and execute all of the instructions with a single stroke of the enter key.

In[37]:= **Clear["@"]**
   a[n_] := {0, −g}
   r[n_] := r[n] = r[n − 1] + h v[n]
   v[n_] := v[n] = v[n − 1] + h a[n − 1]
   r[0] = {0, 0}; v[0] = {30, 40}; g = 9.8; h = 0.1;
   g1 = ListPlot[Table[r[n], {n, 0, 80}], AspectRatio −> Automatic,
       AxesLabel −> {"x", "y"}];

> **AspectRatio->Automatic** makes the *x* and *y* coordinate systems have the same scale.

The first thing we might do is to compare this with the analytic solution, and then proceed to add air resistance. Before going on, we are going to make one more modification to the method we have been using, which changes it to a so-called half-step or leap-frog method and gives us more accuracy. Here is the only modification: We will add the instruction

$$v[1] = v[0] + h/2\ a[0] \tag{7}$$

which has the effect of using the acceleration at the beginning of the first interval to find the velocity *halfway* through the first interval, which we might call $v_{1/2}$. Thereafter we will find $v_{3/2}$, $v_{5/2}$, ..., the velocities at the centers of the time intervals, and these will be used to make the **r**-interval steps. This produces a significant improvement in our numerical analysis technique, less so when the force is velocity dependent than in the case of planetary motion where the force is position dependent. Comparing the old and new approaches is something you might wish to try. Here are the new equations.

```
In[42]:=  Clear[a, r, v]
          a[n_] := {0, -g}
          r[n_] := r[n] = r[n - 1] + h v[n]
          v[n_] := v[n] = v[n - 1] + h a[n - 1]
          r[0] = {0, 0}; v[0] = {30, 40}; g = 9.8; h = 0.1;
          v[1] = v[0] + h/2 a[0];
          g2 = ListPlot[Table[r[n], {n, 0, 82}], AspectRatio -> Automatic,
              PlotStyle -> PointSize[.015], AxesLabel -> {"x", "y"}];
```

We will compare the analytic solution for the position with our numerical analysis approach. The analytic solution is

$$\text{In[47]:=}\quad ra := \left\{30\,t,\ 40\,t - \frac{1}{2}\,9.8\,t^2\right\}$$

This may be graphed with **ParametricPlot**, as follows:

```
In[48]:=  g3 = ParametricPlot[ra, {t, 0, 8.2}, AspectRatio -> Automatic];
```

Now we compare the numerical analysis solution with the analytic solution.

```
In[49]:=  Show[g2, g3];
```

You can see that the comparison is very good.

4.6 Projectile Motion with Air Resistance

We now add drag to our problem, a force that is opposite to the velocity, a situation we illustrate in the next figure. We will assume a quadratic dependence of the force on the velocity, that is,

$$F = -\frac{1}{2} C \rho_a A v^2 \tag{8}$$

where C is the drag coefficient, ρ_a is the density of air, A is the cross section of the object perpendicular to the velocity vector, and v is the magnitude of the velocity. We can use the figure below and similar triangles to show that

$$F_x = \frac{1}{2} C \rho_a A v v_x \text{ and } F_y = -W - \frac{1}{2} C \rho_a A v v_y \tag{9}$$

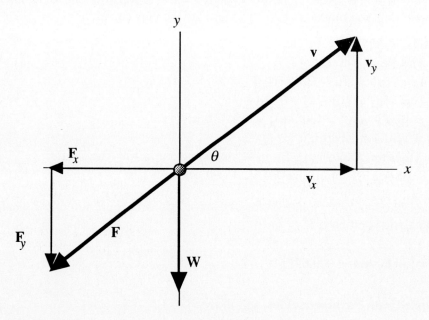

Consequently, the components of the acceleration are:

$$a_x = -kvv_x \text{ and } a_y = -g - kvv_y \tag{10}$$

where

$$k = \frac{1}{2} C \rho_a \frac{A}{m} \tag{11}$$

and m is the mass of the object.

4.6 Projectile Motion with Air Resistance

Problem 9. Use the figure above and similar triangles to derive the formulas for the x and y-components of **F**. Also solve for the acceleration components using Newton's second law.

Problem 10. Assume the object is falling straight down at terminal velocity. In that case it is in equilibrium. Show that the terminal velocity is given by

$$v_T = \sqrt{\frac{2mg}{C\rho_a A}} \qquad (12)$$

You will be able to understand the following set of commands if you recall from a previous section that the magnitude of a vector is the square root of its dot product with itself, and if you recall that the x- and y-components of **v** are **v[[1]]** and **v[[2]]**, respectively. The object modeled with these equations is a baseball, $m = 0.145$ kg, $r = 0.0367$ m, $C = 0.46$, thrown or hit with an initial speed of 50 m/s at an angle of 53 degrees with respect to the horizontal. Try it.

```
In[50]:= Clear["@"]
    magv[v_] := √v.v ;
    a[v_] := {-k magv[v] v[[1]], -g - k magv[v] v[[2]]}
    v[n_] := v[n] = v[n - 1] + h a[v[n - 1]]
    r[n_] := r[n] = r[n - 1] + h v[n];
    v[0] = {30, 40}; r[0] = {0, 0}; g = 9.8; h = .025;
    v[1] = v[0] + h/2 a[v[0]];
    k = 1/2 0.46 1.2 π .0367^2/.145;
    g2 = ListPlot[Table[r[n], {n, 0, 350}], AxesLabel -> {"x", "y"}];
```

Problem 11. Use the model and parameters just given to compare the trajectory of a baseball with and without air resistance. A simple way to obtain the trajectory in a vacuum is to let $k = 0$. By what factor does the range increase without air resistance? Chapman (1968) wrote, "Aerodynamic forces on a baseball are relatively small and have only a small percentage effect on a trajectory." Do you agree with that statement?

Problem 12. The density of air in Denver is about 88% of the density of air in Kansas City. How much farther will the ball described by the preceding model go in Denver than in Kansas City?

Problem 13. The longest home run ever measured was hit by Roy (Dizzy) Carlyle in a minor league game. It traveled 618 ft before it landed outside of the ball park. Assuming its initial velocity made an angle of 40 degrees with respect to the horizontal, what was its initial speed?

Problem 14. Bo Jackson made a flat-footed 300-ft throw (no bounces) from left field to home plate to put out the 10th inning tying run. With what approximate initial speed did he throw the ball?

Problem 15. Assume a Ping-Pong ball ($m = 0.0024$ kg, $r = 0.019$ m, $C = 0.5$) has an initial velocity = {30, 40}, the same as the baseball we have been modeling. What will be its range in air? Explain in words why its range is so much smaller than the range of a baseball thrown with the same velocity in spite of the fact that its cross-sectional area is smaller. Do you think that one single property of an object that has the most profound influence on its range when thrown in air? If so, what is it? Explain.

■ 4.7 Orbital Motion

The force that governs the orbital motion of planets, stars, and galaxies is Newton's universal law of gravitation. The magnitude of the attractive force between a mass m in the vicinity of a mass M is given by

$$F = G \frac{Mm}{R^2} \qquad (13)$$

where G is the gravitational constant and R is the distance between the masses. The figure below shows the salient features of our situation with large mass M, small mass m, and their position vectors. The diagram also illustrates that $\mathbf{R} = \mathbf{r}_2 - \mathbf{r}_1$. In what follows we will assume that the origin of the coordinate system is at the center of mass of this two-body system.

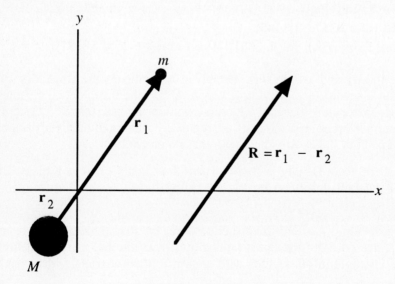

We can use essentially the same Mathematica equations as we did for the motion of a baseball, but you can see that the difficulties are compounded by the fact that we have two

4.7 Orbital Motion

masses to consider rather than one. Thus we must deal with the following two vector equations which express Newton's second law for the objects:

$$m\mathbf{a}_m = -G\frac{Mm}{R^2}\frac{\mathbf{r}_1}{r_1} \quad \text{and} \quad M\mathbf{a}_M = -G\frac{Mm}{R^2}\frac{\mathbf{r}_2}{r_2} \tag{14}$$

However, we may make some significant simplifications in the case that $M \gg m$. This assumption allows us to fix the mass M at the origin of the coordinate system where it is essentially at rest, allowing us to drop the second equation from our consideration and making $\mathbf{R} = \mathbf{r}_2$. If you wish to deal with something like a double-star system where the masses are more nearly equal, then you must deal with both equations. (If you simply wish to find the motion of the mass m relative to the mass M, then you should consult a more advanced text in mechanics where the topic of reduced mass is examined.)

Next, we resolve the various vectors into their x- and y-components and rewrite the equation in Mathematica's vector notation as follows:

In[54]:= $\{\mathbf{a}_x, \mathbf{a}_y\} = -G\frac{M}{R^2}\left\{\frac{x}{R}, \frac{y}{R}\right\}$

where $R^2 = x^2 + y^2$. Recall that in Mathematica notation $\mathbf{x} = \mathbf{R[[1]]}$, $\mathbf{y} = \mathbf{R[[2]]}$, while the magnitude of $\mathbf{R}$ is $\sqrt{\mathbf{R}.\mathbf{R}}$. With those reminders, and noting that we have used a lower-case $\mathbf{r}$, you should be able to understand the following model for orbital motion. For the motion of the Earth about the Sun, we have set $M = 1.991 \cdot 10^{33}$ kg and $R = 1.496 \cdot 10^{11}$ m. Our initial position is $\{R, 0\}$, and our initial velocity is $\{0, v_0\}$ with $v_0 = \sqrt{GM/R}$ which balances the centripetal acceleration GM/R^2 with v_0^2/R, causing the Earth travel in a circular orbit. For h we choose 21,600 s (6 hours). We plot the position of the Earth twice a week for an entire year. Here is our model, which, uses SI units. We begin with a global **Clear**.

In[55]:= **Clear["Global`*"]**

> We use this particular **Clear** because some of our variables, G and M, are capital letters.

In[56]:= $r[0] = \{1.496\ 10^{11}, 0\};\ v[0] = \left\{0, \sqrt{\frac{GM}{1.496\ 10^{11}}}\right\};$
$G = 6.6726\ 10^{-11};\ M = 1.991\ 10^{30};\ h = 21600;$
$v[1] := v[0] + \frac{h}{2} a[0];$
$a[n\_] := -GM/(r[n].r[n])^{\frac{3}{2}} \{r[n][[1]], r[n][[2]]\}$
$v[n\_] := v[n] = v[n-1] + h\ a[n-1]$

```
r[n_] := r[n] = r[n - 1] + h v[n]
ListPlot[Table[r[n], {n, 0, 1461, 21}], AspectRatio -> Automatic,
   AxesLabel -> {"x", "y"}];
```

Problem 16. Give a one-sentence description of each line in the previous set of commands. For example: Line 1 is a statement of the initial conditions of the problem of the Earth moving around the Sun in a circular orbit.

Astronomers and astrophysicists frequently use another set of units for motion around the Sun and the motion of double stars. In this system of units the unit of mass is the mass of the Sun, the unit of distance is the distance between the Earth and the Sun 1 astronomical unit (AU), and the unit of time is 1 year. In these units, which we shall call solar units, $G = 39.6$. The model is as follows:

In[60]:= **Clear["Global`*"]**

$$r[0] = \{1, 0\};\ v[0] = \left\{0, \sqrt{\frac{G\,M}{1}}\right\};\ G = 39.6;\ M = 1;\ h = .0006835\,;$$

```
v[1] := v[0] + h/2 a[0];
a[n_] := -G M/ (r[n].r[n]) ^(3/2) {r[n][[1]], r[n][[2]]}
v[n_] := v[n] = v[n - 1] + h a[n - 1]
r[n_] := r[n] = r[n - 1] + h v[n]
ListPlot[Table[r[n], {n, 0, 1461, 20}], AspectRatio -> Automatic];
```

where we have, once again, chosen an initial velocity appropriate for circular motion.

Problem 17. We would like some checks on our model. In this problem we check to see if the circumference of the model is approximately the circumference of a circle of radius one. In calculus we learn that the arc length of a curve is

$$\int_a^b \sqrt{(x'(t))^2 + (y'(t))^2}\, dt \tag{15}$$

We, on the other hand, do not have analytic expressions for x and y, hence we cannot take their derivatives to find the x- and y-components of the velocity. However, we do have the components of the velocity, so we may approximate the distance the Earth travels in the time interval h as vh. In Mathematica $\mathbf{v} = \sqrt{\mathbf{v[n].v[n]}}$. We add up all of these distances with the **Sum** command. Thus,

In[65]:= $\sum_{n=1}^{1461} \left(\sqrt{v[n].v[n]}\right) h$

Execute this command and compare it to the circumference $2\pi(1)$ predicted for the circumference of a circle of radius 1 AU. What is the percent difference?

4.7 Orbital Motion

Problem 18. Apply one of these models to another system, such as Sun-Mars system. You may want to change both h and the number of points calculated. Note that in the preceding model we have used approximately 1500 points per period of revolution. That is, perhaps, close to the minimum number one should use; more would be better.

Problem 19. The circular orbit of our solar units model is not, perhaps, as interesting as an elliptical orbit with a larger eccentricity, $\epsilon = 0.5$. Let's keep the semi-major axis, $a = 1$, the same, but let the maximum distance be $a(1 + \epsilon) = 1.5a$, and the minimum distance be $a(1 - \epsilon) = 0.5a$. For initial conditions, therefore, choose $\mathbf{r} = \{0.5, 0\}$ and $\mathbf{v} = \{0, 10.8995\}$. The point of closest approach to the Sun is called perhelion, while the most distant point is called aphelion. Where in this orbit is the Earth going the fastest, perhelion or aphelion? How can you tell this from the graph of the orbit? Where is the center of attraction?

Problem 20. Continuing with the orbit graphed in Problem 19, calculate the total energy at each point in the orbit. Energy is conserved in motion under a central force, which is the kind of motion we are modeling. A check on the validity of the model is whether or not it conserves energy. We will discuss energy more completely in a subsequent chapter, but for the moment you may accept on faith that the total energy is given by

$$E = \frac{1}{2} mv^2 - \frac{GMm}{r} \qquad (16)$$

In this expression, the first term is the kinetic energy of the mass m, and the second term is the gravitational potential energy of the system. Since we are only interested in changes in energy, we will divide by m and find the energy per unit Earth mass. In Mathematica this calculation becomes:

In[66]:= **energy[n_] := $\frac{1}{2}$ (v[n].v[n]) – G M / $\sqrt{\text{r[n].r[n]}}$**

Use the results of the solar units model to make a graph of **energy[n]** versus **n** for the entire orbit. Is energy conserved? By what percent does the total energy change with this model? Continue with this problem by halving the value of h we used in the model and doubling the number of points so that you still get a complete orbit. (If computer time is a problem, plot the values for one-half of an orbit.) Now plot the energy again. Does decreasing h result in better conservation of energy?

Problem 21. Another quantity that is conserved in motion under a central force is the angular momentum, and another check on the validity of the model is whether or not it conserves angular momentum. You will learn about angular momentum in your physics course, but for now it is enough to know that the magnitude of the angular momentum per unit Earth-mass in the case we are considering is

In[67]:= **angularmomentum[n_] := r[n][[1]] v[n][[2]] – r[n][[2]] v[n][[1]]**

Use the results of the solar units model to make a plot of the angular momentum versus **n** for all of the points in the orbit. Is the angular momentum constant? By what percent does it vary?

Problem 22. Do you think the distance travelled in traversing this orbit is longer, shorter, or the same as a circular orbit with $r = 1$ au. Calculate the length of the orbit using the formulas in Problem 17.

Problem 23. You may also wish to try to model the orbit of an electron around a proton. In this case the force law is

$$F = -k \frac{Qq}{R^2} \qquad (17)$$

with $Q = q = e = 1.60 \cdot 10^{-19}$, $k = 8.99 \cdot 10^9$, $m_e = 9.11 \cdot 10^{-31}$, all in SI units. Use initial conditions $\mathbf{r} = \{5.3 \cdot 10^{-11}, 0\}$ and $\mathbf{v} = \{0, 2.19 \cdot 10^6\}$. Find a suitable h.

Problem 24. Use the SI units model to make a family of Earth orbits by choosing different values for the initial velocity keeping the initial position constant. Here are some initial velocities to try: $\{0, 7500\}$, $\{0, 15,000\}$, $\{0, 30,000\}$, $\{0, 45,000\}$. You may wish to modify h as well. Which of these initial velocities produce real orbits? What do you expect to happen for sufficiently large v? Can you get the orbit with $v = 45,000$ to close?

■ 4.8 Moon Shots

The purpose of this section is to model the problem of sending a spaceship from Earth to the Moon and back, orbiting around the back side of the Moon to turn around and head back to Earth. Although this may seem like an old problem as far as NASA is concerned, the orbits of some Earth satellites have recently been corrected by sending them around the Moon. In any case, this is the most elaborate calculation we have done so far, and you should participate in it as much as possible. In the figure below, our spaceship is pictured somewhere on its journey between the Earth and the Moon.

4.8 Moon Shots

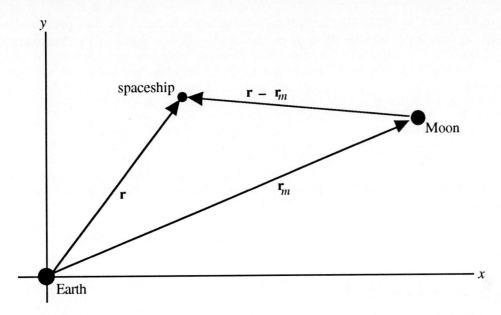

Problem 25. Let M be the mass of the Earth and let m be the mass of the Moon. The spaceship experiences two gravitational accelerations:

$$\mathbf{a}_e = -\frac{GM}{r^2}\frac{\mathbf{r}}{r} \quad \text{and} \quad \mathbf{a}_m = -\frac{Gm}{(|\mathbf{r}-\mathbf{r}_m|)^2}\overline{\mathbf{r}-\frac{\mathbf{r}_m}{|\mathbf{r}-\mathbf{r}_m|}} \qquad (18)$$

Identify each of the symbols in these two formulas. (*Note*: r stands for the magnitude of $\mathbf{r}$, and $|\mathbf{r} - \mathbf{r}_m|$ symbolizes the magnitude of $\mathbf{r} - \mathbf{r}_m$.) Draw the two forces in the preceding figure.

Problem 26. The x-y coordinates assumed for the Earth are $\{0, 0\}$. Assign suitable symbolic Cartesian coordinates for the spaceship and the Moon and express r and $\mathbf{r}_m$ in terms of these coordinates: $\mathbf{r} = \{\underline{\quad}, \underline{\quad}\}$, $\mathbf{r}_m = \{\underline{\quad}, \underline{\quad}\}$, and $\mathbf{r} - \mathbf{r}_m = \{\underline{\quad}, \underline{\quad}\}$.

Problem 27. Note from the two formulas that we need the cube of the magnitude of a vector, for example, r_e^3, which means "find the magnitude of the vector and cube it." First express the magnitude of vectors $\mathbf{r}$ and $\mathbf{r} - \mathbf{r}_m$ in terms of dot products, then cube the results and express the answer as a dot product to the 3/2 power.

Problem 28. Put all of this together to get a formula for the acceleration vector of the spaceship.

Here is the answer you should get

$$\{a_x, a_y\} = \left\{GM\frac{\{x, y\}}{(\mathbf{r}\cdot\mathbf{r})^{3/2}}, -Gm\frac{\{x - x_m, y - y_m\}}{((\mathbf{r} - \mathbf{r}_m)\cdot(\mathbf{r} - \mathbf{r}_m))^{3/2}}\right\} \qquad (19)$$

Problem 29. Now we are ready to write the Mathematica model. We may put the Moon anywhere as long as it is $3.84 \cdot 10^8$ m from the Earth. Before running the Mathematica model below, can you tell where we have put the Moon? Note that we will keep the Moon stationary. Actually it is moving, and in a subsequent problem you get to try to orbit a moving Moon.

Problem 30. Where are we starting the spaceship; that is, what is its initial position relative to the Earth?

The rest of the Mathematica commands have been discussed in previous models. In any case, you should now be in a position to understand the significance of each command in the following model. Finally, the trick is to choose the initial velocity so the spaceship goes around the Moon and returns to the Earth. After many trials and errors we arrived at the following model. Execute the commands and see for yourself. For the record, we give the graph we obtained as well, which also has the position of the Moon plotted.

```
In[68]:= Clear["Global`*"]

        rm = 3.84 10^8 {1, 1}/√2;

        r[0] = 6.37 10^6 {1, 1}/√2;

        me = 5.98 10^24; mm = 7.36 10^22; G = 6.672 10^-11;
        v[0] = {7.605 10^3, 8.090 10^3}; h = 50;
        v[1] := v[0] + h/2 a[0]
        a[n_] := {-r[n][[1]], -r[n][[2]]} G me/(r[n].r[n])^(3/2) +
            {(rm - r[n])[[1]], (rm - r[n])[[2]]}
                G mm/((rm - r[n]).(rm - r[n]))^(3/2)
        r[n_] := r[n] = r[n - 1] + h v[n]
        v[n_] := v[n] = v[n - 1] + h a[n - 1]
        ListPlot[
            Union[{{3.84 10^8, 3.84 10^8}/√2}, Table[r[n], {n, 0, 13000, 75}]],
            AspectRatio -> Automatic, PlotRange -> All,
            AxesLabel -> {TraditionalForm[x], TraditionalForm[y]},
            Prolog -> PointSize[0.005]];
```

4.8 Moon Shots

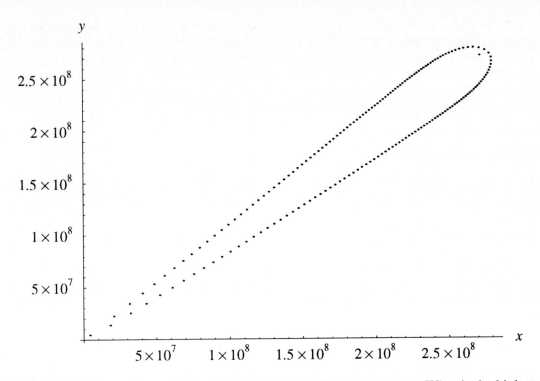

Problem 31. Interpret the variable spacing between dots in this figure. What is the highest speed of the spacecraft? The smallest? How close does the spaceship come to the Moon? If you have lots of time to spend, try finding an orbit that approaches the Moon more closely.

The *Apollo* mission problem illustrates a grave difficulty with our numerical integration method. Near the Moon the acceleration of the spaceship is large and, to get sufficient accuracy, h should be small compared to what it might be in the much larger distance (and time) interval where the spaceship is traveling between the Moon and the Earth. We must choose an excessive number of points to get reasonable accuracy in the vicinity of the Moon. Some kind of adaptive algorithm that allows the size of h to change depending on the acceleration would be much more desirable. In short, we are close to the limit of usefulness for the computational techniques described in this chapter.

All is not lost, however; Mathematica has a command called **NDSolve** that does the numerical integrations we have been doing in the past several chapters, and does it much much better. In later chapters we will make much use of this command.

Problem 32. Here is an equally difficult problem for you. Let the Moon move in a circular orbit described by

$$r_m = \{x_m, y_m\} = r_m \left\{\cos\left(\frac{2\pi nh}{T} - \phi\right), \sin\left(\frac{2\pi nh}{T} - \phi\right)\right\} \qquad (20)$$

where T is the period of the Moon (approximately 30 days) and ϕ is an arbitrary phase angle to start the Moon at whatever initial position you desire. Now adjust the initial velocity of the spaceship so that it goes to the moving Moon and returns to the Earth. This may be a long and tedious problem. It is included to illustrate that the level of complexity goes up as fewer and fewer simplifying assumptions are made about the real situation we wish to model, that is, as more complexities are introduced.

Mostly Mathematica

1. If **a** = 3**i** + 2**j** + **k** and **b** = –2**i** + 4**j** – 3**k**, find: **a** + **b**, **a** – **b**, | **a** | (the magnitude or norm of **a**), | **a** – **b** |, **a.b**, and **a x b**. If necessary, use Mathematica's **??Cross** to find information about the cross product.

2. Find the angle between **a** and **b**.

3. One particle moves along the path $y = x$. Another particle travels around a circle of radius 5 centered at the origin. Where do these paths cross? This problem should refresh your memory about Mathematica's **Solve** command.

Explorations

1. Will other force laws give closed orbits? For example, suppose the force of attraction is

$$\mathbf{F} = -\frac{GMm}{r}\mathbf{r} \tag{21}$$

To study some of these other possibilities, it pays to further simplify the problem. Assume we live in a universe with $G = 1$ and $M = 1$. Start the orbiting mass at $\{1, 0\}$ with a velocity of $\{0, 1\}$. Begin with a circular orbit, but try some different initial velocities. Here is your (much simplified) model:

```
In[74]:= Clear["@"]
    r[0] = {1, 0}; v[0] = {0, 1}; h = .01;
    v[1] := v[0] + h/2 a[0];
    a[n_] := – {r[n][[1]], r[n][[2]]} / √r[n].r[n]
    v[n_] := v[n] = v[n – 1] + h a[n – 1]
    r[n_] := r[n] = r[n – 1] + h v[n]
    ListPlot[Table[r[n], {n, 0, 6280, 5}], AspectRatio –> Automatic];
```

Also try attractive forces of

$$\mathbf{F} = -GMm\mathbf{r} \quad \text{and} \quad \mathbf{F} = -\frac{GMm}{r}\frac{\mathbf{r}}{r} \tag{22}$$

4.8 Moon Shots

Can you get closed circular and noncircular orbits for these force laws? If so, are the orbits elliptical? For example, with the force law $\mathbf{F} = -GM\mathbf{r}$ can you get elliptical orbits, and if so, does the force point toward a focus as it does for an inverse square force law, or does the attractive force point towards the center?

2. Model the motion of a binary star system. It's best to use solar units or an adaptation of our very simple model. Plot the motions of both stars and also the motion of one star relative to the other. You might start with $m = \frac{1}{4}M$. Add a planet to the binary star system. Are there stable orbits for this planet that takes it around both stars?

3. Consider the motion of a satellite about a planet with nonspherical mass distribution such as a large mountain on an otherwise featureless surface. How does the orbit evolve with time?

4. Model the motion of a projectile on the surface of the Earth where the projectile goes high enough so that the acceleration of gravity cannot be assumed to be constant. Include air resistance as a complication. Also include the change in the density of air with altitude. How might the effect of wind be taken into account?

5. Model the motion of Halley's Comet. This model will make significant, perhaps excessive, demands on both computer time and memory. The difficulty lies, once again, in the fact that h must be small when the comet is near the Sun where the comet spends only a small fraction of its time. You will need the approximate parameters of its orbit. The eccentricity ϵ is 0.96728, the semimajor axis a is 17.94 AU, and the energy at any point in the orbit is

$$\frac{1}{2}mv^2 - \frac{GMm}{r} = -\frac{GMm}{2a} \qquad (23)$$

6. Model the orbital and radial motion of a particle under the influence of the central force law

$$F = \frac{5}{r^5} - \frac{3}{r^3} \qquad (24)$$

7. Simulate the hunter and the monkey demonstration. A hunter shoots a gun aimed directly at a monkey hanging from a limb of a tree some distance away. At the instant the hunter fires the gun the monkey drops from the limb. Does the bullet hit the monkey?

Reference

Chapman, S. Catching a baseball, *Am. J. Physics* 36, (1968): 868.

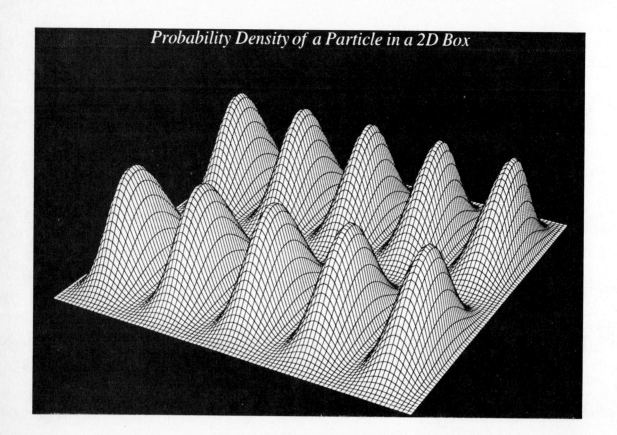

Probability Density of a Particle in a 2D Box

CHAPTER 5 Using Mathematica to Do Traditional Physics Problems

■ 5.1 Introduction

There are many problems in physics that require the solution of a system of one or more equations. This is a stage in the solution of a problem where it is easy to make an error and get the problem wrong. The more equations in the system, the more likely it is to make a mistake and get an incorrect answer. In this chapter we look at a variety of problems and we use Mathematica to do the mathematics. We are not looking for new insights, we are merely trying to improve our ability to get correct answers once we understand the physics of the problem. Although we give several practice exercises, you can find many more example problems and exercises in your physics textbook that you can practice solving with Mathematica. Remember, understanding the physics of the problem (drawing a free-body diagram, identifying the forces, and setting up the equations) will always be *your* task:

5.2 Kinematics Problems

Mathematica cannot set up any problem, it can only solve problems once they are set up by a human being.

■ 5.2 Kinematics Problems

Example Problem 1. A car traveling at 60 km/h is 1 km behind another car traveling at 70 km/h. How long will it take for the first car to overtake the second, if the first car can accelerate at 3 m/s/s and the second car continues to move at constant speed?

Perhaps the first thing we should recognize is that it is impossible for any real car to continue to accelerate. Because its power is limited, a car will reach some limiting velocity. Nevertheless, a physics problem does not have to correspond to the real world, so we shall go ahead and solve this problem.

Remember, you first have to know how to solve the problem yourself before attempting to do anything with Mathematica. Begin by introducing a coordinate system in this problem. The *x*-axis lies along the line connecting the two cars with its origin at the position of the first car, which we may call car 1 while the second car will be called car 2, at the instant, $t = 0$, when the problem begins. It would be a good idea to represent this problem with a picture, car 1 at the origin and car 2 1 km to the right on the *x*-axis.

First describe the strategy in normal mathematics. The positions of car 1 and car 2 as a function of time are

$$x_1 = 60 \frac{1000}{3600} t + \frac{1}{2} 3 t^2, \quad x_2 = 1000 + 70 \frac{1000}{3600} t \qquad (1)$$

Note that we have converted km/h to m/s. When car 1 overtakes car 2, the two positions are identical; thus our method will be to solve these two equations simultaneously for the common position and the time. In Mathematica we write

In[1]:= x1 = 60 $\frac{1000}{3600}$ t + $\frac{1}{2}$ 3 t^2

In[2]:= x2 = 1000 + 70 $\frac{1000}{3600}$ t

Now we ask for an approximate value for the time when these two positions are the same.

In[3]:= **Solve[x1 == x2, t] // N**

We are not interested in the negative solution because it refers to a time before the events under consideration happen, so the positive time is the desired solution. To find the position of either car at this instant we find **x1** given that **t** = 26.7624. We do that with the /. command. Thus

In[4]:= **x1 /. t —> 26.7624**

To satisfy our own curosity about the reasonablness of this problem, we also find the speed of car 1 at this time. We make use of the fact that

In[5]:= **v1 = ∂_t x1**

In[6]:= **v1 /. t —> 26.7624**

Converted to km/h, this is

In[7]:= **96.9539 $\dfrac{3600}{1000}$**

which is quite unreasonable unless we have a rocket sled on the salt flats in Utah.

Example Problem 2. Suppose in the previous example that car 1 can achieve a maximum velocity of 150 km/h. In that case, where and when does car 1 overtake car 2?

We begin by finding out when the speed of car 1 reaches 120 km/h.

In[8]:= **Solve$\left[\text{v1}\,\dfrac{3600}{1000} == 120., t\right]$**

Next, we separate the problem into two parts, before car 1 reaches 120 km/hr and after that instant. Find out where the two cars are when car 1 reaches 120.

In[9]:= **x1 /. t —> 5.5556**
       x2 /. t —> 5.5556

It's clear that the first car hasn't gained that much. In any case, we now rewrite the position equations for times later than 5.5556.

In[11]:= **x1 = 138.89 + 120. $\dfrac{10}{36}$ t**

x2 = 1108.03 + 70. $\dfrac{10}{36}$ t

Now we find when these two positions are the same.

In[13]:= **Solve[x1 == x2, t]**

The total time is, therefore,

In[14]:= **69.7781 + 5.5556**

And, the position is

In[15]:= **x1 /. t –> 69.7781**

This completes the solution to the problem.

Problem 1. A train is traveling at 100 km/h when the engineer sees another train 1 km in front on the same track and radar shows this train is traveltravelinging at 60 km/h. If the overtaking train can slow down at 0.05 m/s/s, will a collision between the trains be avoided? If not, when will it take place? If so, how close do the trains get?

Problem 2. This is a bit tricky, but suppose the engineer on the overtaking train has some control over the deceleration of the train. What minimum deceleration will avoid a collision?

Example Problem 3. An airplane that can fly 200 m/s in still air has a heading of 30 degrees. All angles are measured clockwise from North. The wind is blowing toward the southwest with a speed of 25 m/s. With what speed and in what direction does the plane move relative to the ground (ground track)?

We will need vectors to solve this problem, and we need a well-known relationship between relative velocities, namely

$$\mathbf{v}_{pg} = \mathbf{v}_{pa} + \mathbf{v}_{ag} \qquad (2)$$

This may be read as follows: The velocity of the plane relative to the ground is the sum of the velocity of the plane relative to the air and the velocity of the air relative to the ground (wind). Notice the order of the subscripts.

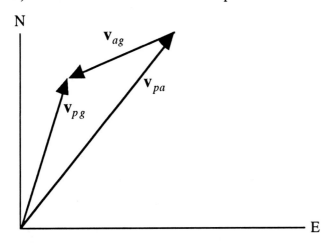

The first step is to sketch a vector diagram showing the *x*- and *y*-components of these vectors, a sketch which is shown above. Like any vector sum, note that this sum can be represented as a triangle using the definition of vector addition.

In[16]:= **Clear["@"]**

We express the vectors in Mathematica.

In[17]:= **vpa = 200. {Sin[30 Degree], Cos[30 Degree]}**

In[18]:= **vag = 25. {−Cos[45 Degree], −Sin[45 Degree]}**

In[19]:= **vpg = vpa + vag**

The magnitude of this vector is

In[20]:= $\sqrt{\text{vpg} \cdot \text{vpg}}$

In what direction does this vector point? The angle between North and the vector is

In[21]:= **ArcTan[vpg[[1]] / vpg[[2]]] / Degree**

Example Problem 4. The usual problem for a pilot is to choose the heading of the ground track and find the appropriate direction to aim the airplane as well as the ground speed. Suppose a pilot wishes to have a ground track heading of 60 degrees. The wind is currently 20 m/s out of the southwest. If the air speed of the airplane is 80 m/s, what is the ground speed and the heading of the plane?

We begin by drawing a vector diagram, roughly to scale.

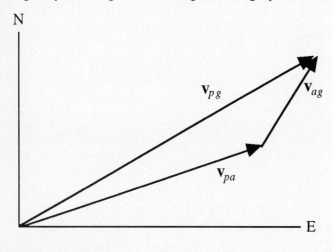

In[22]:= **Clear["@"]**

5.2 Kinematics Problems

Here is the wind.

    In[23]:= **vag = 20. {Sin[45 Degree], Cos[45 Degree]}**

Here is the velocity of the plane relative to the ground. We know what direction we want, but the magnitude of the velocity (**mag**) is not known.

    In[24]:= **vpg = mag {Sin[60 Degree], Cos[60 Degree]}**

We know the plane can fly at 80 m/s relative to the air, and we want to know the direction, x.

    In[25]:= **vpa = 80. {Sin[x], Cos[x]}**

What follows is the vector equation for the relative velocities.

    In[26]:= **eqn = vpg == vpa + vag**

Notice that we have two unknowns, **mag** and **x**. We attempt to solve for both unknowns.

    In[27]:= **Solve[eqn, {mag, x}]**

Our solution set is empty! That is surprising, since we are quite sure there is a solution. We also get a warning that some solutions may not be found. In fact, our solution is not found. Here is where life gets tricky. We try to separate out the two unknowns using Mathematica's **Eliminate** command. Let's eliminate **mag**.

    In[28]:= **neweqn = Eliminate[eqn, mag]**

and then solve for x.

    In[29]:= **Solve[neweqn, x]**

Referring to our figure, we believe that

    In[30]:= **1.11195 / Degree**

is the correct solution.

We can also eliminate x.

    In[31]:= **eqnthree = Eliminate[eqn, x]**

    In[32]:= **Solve[eqnthree, mag]**

Our ground speed is approximately 99 m/s, and our heading should be approximately 64 degrees.

Problem 3. A plane that can fly 100 m/s in still air is to travel to a town with a heading of 330 degrees from the current position of the airplane. A 15-m/s wind is blowing from the northeast. What heading should the plane have? What is its groundspeed in that case?

■ 5.3 Statics Problems

In[33]:= **Clear["Global`*"]**

Example Problem 5. A 2-m long beam is supported at its left end by a pin and by its right end by a wire that goes back toward the wall that supports the beam. The wire makes and angle of 37 degrees with respect to he horizontal. The beam weighs 200 N. What is the tension in the wire and the horizontal and vertical reactions at the pin?

Begin by sketching a free-body diagram of the beam.

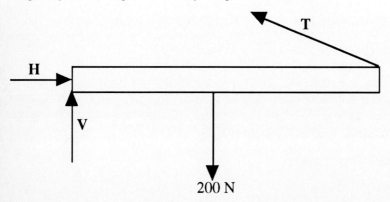

Since the beam is in equilibrium, we must sum the horizontal and vertical forces as well as the torques about some point to zero. Call the horizontal and vertical reactions at the left end H and V, respectively, and call the tension in the wire T. The sum of the forces in the x-direction is

In[34]:= **xeqn = H − T Cos[37 Degree] == 0**

In the y-direction we have

In[35]:= **yeqn = V + T Sin[37 Degree] − 200 == 0**

We assume the 200-N weight acts in the center. Taking torques about the left end gives

In[36]:= **torque = −200 1 + T Sin[37 Degree] 2 == 0**

We solve for the unknowns **T**, **H**, and **V**.

In[37]:= **Solve[{xeqn, yeqn, torque}, {H, T, V}] // N**

Problem 4. A gate weighing 450 N is 3.5 m long and 1.5 m high. It is supported by two hinges, one at the bottom of the gate and the other at the top. The bottom hinge can provide both a vertical force and a horizontal force, but the top hinge can only provide a vertical force. Find all three of these forces.

Example Problem 6. A structure as shown in the following figure allows a cart carrying a weight W to move back and forth with x varying from 0.1 m to 2.4 m. The structure is supported by a bearing on the floor that provides both a vertical component V and a horizontal component H. At the top, a cable provides a tension T. If the weight of the structure is quite small compared to the weight W of the load, find how T, V, and H vary with x and W. If $W = 1000$ N, what is the maximum tension in the cable? Most of the details of the structure are shown in the figure.

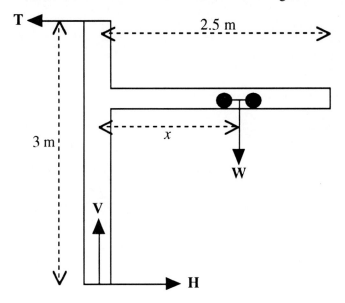

In[38]:= **Clear["Global`*"]**

Adding the horizontal forces gives

In[39]:= **xeqn = H − T == 0**

Adding the vertical forces gives

In[40]:= **yeqn = V − W == 0**

The torque about the bearing at the bottom is

In[41]:= **torque = T 3 − W x == 0**

It's easy to find T in terms of the weight and the geometric factors:

In[42]:= **Solve[torque, T]**

The maximum tension occurs when x is a maximum, so the maximum tension is

In[43]:= **1000 2.4 / 3**

Given the horizontal force equation we can write $H = Wx/3$. Finally, V does not change with x, as we can see from the vertical force equation.

Problem 5. A ladder with a mass of 10 kg and which is 4 m long leans against a smooth wall. The bottom of the ladder rests on a floor 1 m from the wall, and the coefficient of friction between the ladder and the floor is 0.7. What are the horizontal and vertical reactions to the ladder by the floor, and what is the horizontal reaction to the ladder by the wall?

■ 5.4 Dynamics Problems

In[44]:= **Clear["Global`*"]**

Example Problem 7. Here is a simple Atwood's machine problem as illustrated in the diagram below. Assuming the pulley has no mass, what are the accelerations of the two masses.?

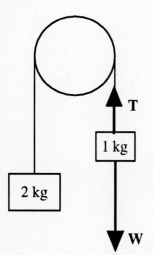

You should begin by drawing a free-body diagram for each mass. Applying Newton's second law to the 1-kg mass gives the first equation

5.4 Dynamics Problems

In[45]:= **eqn1 = T – 1 g == 1 a**

where *T* is the tension in the connecting rope.

The equation of motion for the second mass is

In[46]:= **eqn2 = 2 g – T == 2 a**

We solve these two equations to obtain the solution.

In[47]:= **Solve[{eqn1, eqn2}, {T, a}]**

Problem 6. A 1-kg mass rests on a frictionless incline which makes an angle of 30 degrees with respect to the horizontal. Find the acceleration of the mass down the plane and the normal force of the plane on the mass. Set this up as two equations in two unknowns and then use the **Solve** command.

Example Problem 8. In Example Problem 7, the pulley has a moment of inertia *I* and a radius *R*. Once again, find the acceleration of the masses and the angular acceleration of the pulley. In this case, solve the problem initially in terms of the parameters. Assume the mass on the right is **m1** and the mass on the left is **m2**.

We keep the same two equations we had earlier, but we add two more, noting that the tensions on either side of the pulley are different. Here are the first two equations modified for this change.

In[48]:= **Clear["Global`*"]**
 eqn1 = T1 – m1 g == 1 a
 eqn2 = m2 g – T2 == 2 a

Here are the next two. They come from rotational dynamics

In[51]:= **eqn3 = (T2 – T1) R == i α**
 eqn4 = a == R α

We cannot use **I** for the moment of inertia because in Mathematica **I** is reserved for $\sqrt{-1}$.

In[53]:= **sol = Solve[{eqn1, eqn2, eqn3, eqn4}, {T1, T2, a, α}]**

This solution is pretty abstract: let's assign some values. Let **R = 0.5** , **i = 1/8**, **g = 9.8**, **m1 = 1**, **m2 = 2**. Here is one way to do that.

In[54]:= **sol /. {R –> .5, i –> 1/8, g –> 9.8, m1 –> 1, m2 –> 2}**

When confronted by a system like this, before solving it my students will invariably predict the tension in the right-hand string to be equal to the weight of **m2**, 19.6. I suppose they see the string holding up the weight. To look at this difficulty in more detail, take **R** = 0.5, i = 1/8, **g** = 9.8, and **m2** = 2 (as above), then graph the tension **T2** as a function of **m1**.

In[55]:= **Plot[T2 /. sol /. {R –> 0.5, i –> 1/8, g –> 9.8, m2 –> 2}, {m1, 0, 5},
        AxesLabel –> {"m1", "T2"}, PlotRange –> {0, 30}];**

Notice from the graph that when **m1 = m2 = 2**, the tension **T2** is 19.6. But that is when the system is in equilibrium. For all values of **m1** larger than 2, the tension **T2** is larger than the weight of **m2** and it accelerates upward. For all values of **m1** smaller than 2, the tension **T2** is smaller than the weight of **m2**, and it accelerates downward. Once you think about it, that makes sense. We could also kill a mouse with an elephant gun and make our point this way:

In[56]:= **Limit[T2 /. sol /. {R –> 0.5, i –> 1/8, g –> 9.8}, m1 –> m2]**

This result informs us that **T2** approaches the weight of **m2** as **m1** approaches **m2**, in other words, as the system approaches equilibrium.

Problem 7. A rope connecting two different masses hangs over two identical pulleys of radius R and moment of inertia I. The pulleys are supported on frictionless axles. Find the acceleration of the masses, the angular acceleration of the pulleys, and the tension in each of the three sections of the rope as a function of the two masses, R, I, and g. The picture is shown below. Begin by making free-body diagrams for each of the masses and pulleys. Under what circumstances is the tension in the middle section of the rope equal to the weight of either of the masses?

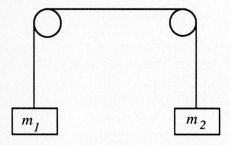

■ 5.5 Energy Problems

In[57]:= **Clear["Global`*"]**

Example Problem 9. A bungee cord behaves like a spring and the force it produces is $F = -kx$ where x is the displacement from its unstretched length L. A bungee cord is connected

5.5 Energy Problems

at one end to a bridge and to the other end to Joe whose mass is m. What should k be so that when Joe jumps with the bungee cord originally unstretched he does not hit the river on his first big bounce? The river is a distance h below the bridge.

Our plan of attack is to find k so that Joe just falls a distance h. A larger k will ensure that he does not hit the river. You should probably draw a picture showing the bridge, Joe, as a point mass, after falling a distance L, and Joe, still as a point mass, at the bottom of his plunge at a distance h from the bridge. The bungee cord only begins to stretch when Joe has fallen a distance L. At this point Joe has lost mgL joules of gravitational potential energy, and he has gained kinetic energy. He now continues to drop an additional distance x, stretching the cord all the while, increasing the elastic energy of the Joe-bungee cord system to $(kx)^2/2$. We write

In[58]:=
$$\textbf{potentialenergylost} = \textbf{m g L} + \textbf{m g x}$$

In[59]:= $\textbf{elasticenergygained} = \dfrac{1}{2} \textbf{k x}^2$

We know that at the bottom all of Joe's potential energy turns into elastic energy, and we also know that $h = L + x$ for Joe to just hit the water. So we solve these two equations as follows:

In[60]:= **solution =**
  Solve[{potentialenergylost == elasticenergygained, h == L + x}, {k, x}]

Suppose $L = 10$ m, $h = 30$ m, Joe has a mass of 75 kg, and $g = 9.8$ m/s/s. What should the value of k be?

In[61]:= **solution /. {L -> 10, h -> 30, m -> 75, g -> 9.8}**

Problem 8. Suppose in the problem we just solved that we want a safety margin of 10 m; that is, no part of Joe should come to rest less than 10 m above the river. Also supposewe want to take into the fact that Joe is not a point mass, rather he is 1.5 m tall, and the bungee cord is attached to his ankles. What should k be in this case? Where does Joe come to rest after all the bouncing is finished?

Problem 9. A rollercoaster is moving at 3 m/s when it tops a hill 52 m above the ground. The next hill, after going through a valley, is 57 m above the ground. Will the rollercoaster make it over the next hill? If so, how fast will it be going at the top? If not, how high does it get before it comes to a stop? Assume the rollercoaster has no rolling friction or air resistance.

Example Problem 10. A mass $m = 1$ kg moves to the right on the x-axis with velocity $v = 2$ m/s. As it passes the origin, it encounters a mysterious force whose potential energy function is

$$U(x) = 5(x - \sin(x)) \qquad (3)$$

To what value of x does the mass move before stopping?

Our strategy will be to use conservation of energy. When the mass stops, all of its kinetic energy will be in the potential energy of the system. We begin by writing the potential energy function.

In[62]:= **Clear["Global`*"]**
       U[x_] = 5 (x − Sin[x])

Next, we try to equate the kinetic energy and the kinetic energy and solve for x.

In[64]:= **solution = Solve$\left[\frac{1}{2} m v^2 == U[x], x\right]$**

What you get is a message from Mathematica that basically admits defeat in solving this equation. In fact, you cannot solve this equation using traditional algebra or trigonometry. However, Mathematica has another command, called **FindRoot**, that may help us. We need to specify all parameters such as m and v in this problem, and we need some kind of guess at the solution.

A good way to arrive at a guess is to graph both sides of the equation and see where the graphs cross.

In[65]:= **Plot$\left[\left\{\frac{1}{2} 1\, 2^2, U[x]\right\}, \{x, 0, 3\}\right]$;**

Since we haven't used **FindRoot** yet, we should inquire about the appropriate syntax. We do this with the

In[66]:= **? FindRoot**

command. Finally, we use the **FindRoot** command, as follows:

In[67]:= **FindRoot$\left[\frac{1}{2} 1\, 2^2 == U[x], \{x, 1\}\right]$**

and we obtain our answer.

Problem 10. Suppose in the previous problem that the mass is 2 kg and the velocity is 4 m/s. To what value of x does the mass move before coming to a stop?

Problem 11. A particle of mass m slides down a frictionless hill that has the shape

$$h(x) = e^{-x} \qquad (4)$$

where $h(x)$ is the height of the hill at position x. If the particle starts at $x = 0$, to what value of x must it slide to acquire a speed of 3 m/s? Assume $g = 9.8$ m/s/s and recall that the gravitational potential energy is *mgh*. Be sure to use a **Clear** before using the same variables we have used above. Then use both **Solve** and **FindRoot** to attempt to solve this problem.

■ 5.6 Momentum Problems in One Dimension

Momentum problems are usually divided into those involving elastic collisions, in which case kinetic energy as well as momentum is conserved, and those involving inelastic collisions where only momentum is conserved. We begin with the latter, and all motion will be assumed to take place along the x-axis.

5.6.1 Inelastic Collisions

In[68]:= **Clear["@"]**

We consider a mass **m1** traveling with initial velocity **v1i** striking a mass **m2** traveling with velocity **v2i** and we are interested in their velocities, **v1f** and **v2f**, respectively, after the head-on collision. Regardless of whether the collision is elastic or inelastic we have the conservation of momentum equation. Here it is:

In[69]:=
> momentum = m1 v1i + m2 v2i == m1 v1f + m2 v2f

Clearly we have a single equation here and the possibility of one or more unknowns. Typically we know the masses and their initial velocities; that leaves the final velocities as unknowns, and there are two such velocities. We can go no further without additional information.

> *Case I.* The object strikes another object at rest and the objects stick together after the collision.

In this case, **v2i = 0** and **v2f = v1f**. In Mathematica we use the given that notation (**/.**) and the set of rules that assign these values in the momentum equation, as follows:

In[70]:= **solution = Solve[momentum /. {v2f -> v1f, v2i -> 0}, v1f]**

Suppose we have an object of very small mass, such as a rifle bullet whose mass is 2 g, going very rapidly, say 500 m/s, which strikes a block of wood with a mass of 5 kg and lodges in it. What will be the final velocity of the block/bullet? Here is how to find the answer.

In[71]:= **v1f /. solution /. {m1 –> .002, m2 –> 5, v1i –> 500}**

> Many students find the *given that, /.*, notation difficult. An alternative approach is to assign variables with = signs.

Related to the problem we just solved, we could first assign the values:

In[72]:= **m1 = .002**
m2 = 5
v1i = 500

and then find **v1f** as follows:

In[75]:= **v1f /. solution**

To my mind this approach is less than sophisticated because now we must **Clear** these three variables before using them again. Think seriously about using the replacement rule (/.) approach.

Problem 12. Find the initial velocity of a 1-g bullet that strikes a 0.5-kg block, initially at rest, and is imbedded in it. After the collision the block/bullet travel with a speed of 0.5 m/s.

| *Case II.* All other situations.

In all other cases we need additional information about the velocity of one of the masses after the collision before we can solve for the other unknown. Here is a typical problem.

Example Problem 11. A small projectile with mass 2 g is traveling at 500 m/s and strikes a 1-kg apple initially at rest. After going through the apple, it continues onward at 300 m/s. How fast does the apple move after the bullet leaves it?

We begin with the same conservation of momentum equation:

In[76]:= **Clear["@"]**
momentumeqn = m1 v1i + m2 v2i == m1 v2f + m2 v2f

We solve this for the velocity of the apple after the collision.

In[78]:= **solution = Solve[momentumeqn, v2f]**

Finally, we substitute in our numbers.

5.6 Momentum Problems in One Dimension

In[79]:= **v2f /. solution /. {m1 -> .002, m2 -> 1, v1i -> 500, v1f -> 300, v2i -> 0}**

5.6.2 Elastic Collisions

In[80]:= **Clear["@"]**

Example Problem 12. A mass **m1** with an initial velocity **v1i** strikes a mass **m2** head on. Mass **m2** has an initial velocity **v2i**. The collision is elastic. Find the velocities of **m1** and **m2** after the collision, **v1f** and **v2f**, respectively.

Conservation of momentum gives

In[81]:= **momeqn = m1 v1i + m2 v2i == m1 v1f + m2 v2f**

Conservation of kinetic energy gives

In[82]:= **energyeqn = $\frac{1}{2}$ m1 v1i^2 + $\frac{1}{2}$ m2 v2i^2 == $\frac{1}{2}$ m1 v1f^2 + $\frac{1}{2}$ m2 v2f^2**

In[83]:= **solution = Solve[{momeqn, energyeqn}, {v1f, v2f}]**

Notice there are two solution sets. What does the first one mean? We are interested in the second solution. We can extract that as follows:

In[84]:= **oursol = solution[[2]]**

These results are a bit opaque without some simplifying assumptions. Sometimes it is informative to look at some individual cases to see if the results agree with our intuition.

Case I. A large mass strikes a small mass at rest.

In[85]:= **Limit[{v1f, v2f} /. oursol /. v2i -> 0, m1 -> ∞]**

Clearly, the large mass continues on almost unimpeded, while the small mass moves off with twice the speed of the colliding mass.

Case II. A small mass strikes a large mass at rest.

Problem 13. You handle this case. Take the limit as **m1 -> 0**. Write out your conclusion.

Case III. A mass strikes a particle of the same mass at rest.

Problem 14. You handle this case. What limit must you take? Write out your conclusion. Does this result agree with the intuition you developed during the misspent part of your life in a pool hall? Can you do this without taking a limit? Try it.

Case III. Two identical masses traveling in opposite directions at the same speed strike each other.

Problem 15. You can also handle this case. What do you conclude?

Mostly Mathematica

1. Execute the commands **??Prime**, **??Table**, and **??ListPlot** to see the syntax of each. Make a table of the first 50 prime numbers using Mathematica's **Table** command. Nest **Prime** in **Table** in **ListPlot** to make a graph of the first 50 prime numbers with a single command.

2. A mouse travels around a circle. The parametric equations of a circle are

$$x(t) = r\cos(t), \quad y(t) = r\sin(t) \qquad (5)$$

Make a **ParametricPlot** of these two equations that generate an entire circle with $r = 2$.

3. Find $v_x(t)$ and $v_y(t)$ by taking the derivatives of x and y in the previous problem. Make a **ParametricPlot** of these two velocity components. What does it tell you about the magnitude of the mouse's velocity? Find both components of the acceleration and the magnitude of the acceleration.

Exploration

1. Develop the Mathematica skills you need to solve momentum problems in two dimensions. Refer to the last chapter for a discussion of vectors in two dimensions.

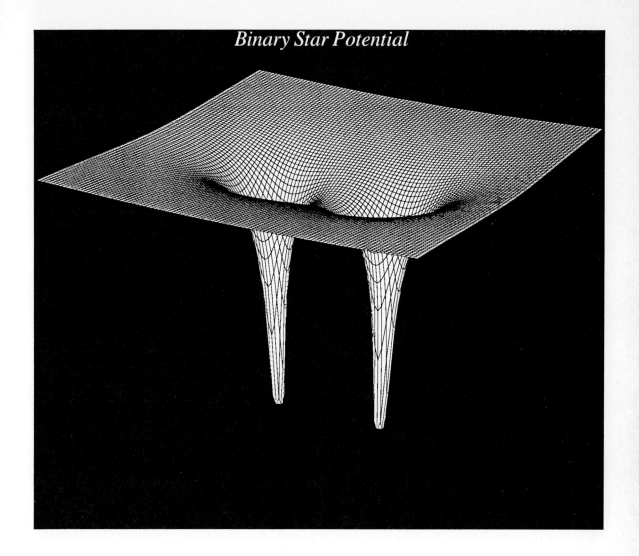
Binary Star Potential

CHAPTER 6 Potential Energy and Conservative Forces

■ 6.1 Introduction

In a standard physics textbook we learn to look at the world from the perspective of Newton's laws, and we see a world governed by forces. From these forces you can find accelerations and, by suitable integrations, velocities and positions. Another, not mutually exclusive, perspective is to look at the world from the point of view of energy. The observation that particles move downhill is essentially the same as the postulate that particles in a system seek a minimum in the potential energy of the system. With this fact and the conservation of energy, it is possible to find the subsequent velocities and positions of the particles in the system. For some problems you can dispense with the concept of force, but we will not choose to go to that extreme. In any case, the perspective of this chapter will be: predict and understand the behavior of a system given the potential energy function. The graphing capabilities of Mathematica will be exploited to help us understand some simple systems. To begin, let's review some of the facts related to potential energy that you should have studied but may have forgotten.

■ 6.2 Potential Energy and Force

There is a definite relationship between the potential energy of a particle and a conservative force acting on it. We define potential energy difference, $U_f - U_i$, as follows:

$$U_f - U_i = -\int_{s_i}^{s_f} \mathbf{F} \cdot \mathbf{ds} \qquad (1)$$

This defining equation is one relationship between $\mathbf{F}$, a conservative force acting on some particle of mass m, and the potential energy U of the system. For example, $\mathbf{F}$ might be the constant force of gravity acting on a particle in the vicinity of the Earth's surface, or it might be the nonconstant force of a spring on a mass. Remember, $\mathbf{F}$ is a conservative force, which excludes friction forces. The initial position of the particle is s_i, and s_f is its final position. This definition of potential energy is by no means a simple one; the integral as written is quite symbolic and, for example, cannot simply be integrated without assigning a coordinate system and expressing $\mathbf{F}$ and $d\mathbf{s}$ in that coordinate system.

We can illustrate with the most common example, namely a particle of mass m in the gravitational field of the Earth. First, we establish a coordinate system. The y-axis will be taken to be in the vertical direction with the origin at the surface of the Earth. With this assumption $\mathbf{F} = -mg\mathbf{j}$ and $\mathbf{ds} = dy\mathbf{j}$. The dot product is, therefore, $-mg\, dy$. Only now can we integrate;

$$U_f - U_i = \int_{y_i}^{y_f} mg\, dy = mgy_f - mgy_i \qquad (2)$$

You should be familiar with this result. Because we are only interested in differences in potential energy, we are free to choose a zero at any value of y. (This is analogous to making a length measurement, $x_f - x_i$, with a ruler, in which case the measurement of length is independent of where the end of the ruler is.) In our example we will take $U_i = 0$ at $y_i = 0$. We also drop the f subscript to obtain

6.2 Potential Energy and Force

$$U = mgy \tag{3}$$

Thus, associated with the force of gravity near the surface of the Earth is the potential energy function $U(y) = mgy$.

If the potential energy function were complicated, we could use Mathematica to graph it. First we would define the potential energy function:

In[1]:= **U[y_] := m g y**

Since Mathematica cannot graph a function with symbolic parameters, we merely assign some simple ones. Thus

In[2]:= **m = 1; g = 9.8;**

and then proceed to graph the function as follows:

In[3]:= **g1 = Plot[U[y], {y, 0, 10}, AxesLabel –> {"y", "U(y)"}];**

To understand the behavior of the system, imagine placing a particle at some nonzero y. The particle will move in such a way as to minimize its potential energy. Thus the particle moves to smaller y, gaining kinetic energy as it goes.

Referring to Equation (1) again, associated with every integral relationship is a derivative relationship. In particular, the derivative relationship associated with the definition of the potential energy is

$$F_x = -\frac{\partial U}{\partial x}, \quad F_y = -\frac{\partial U}{\partial y}, \quad F_z = -\frac{\partial U}{\partial z} \tag{4}$$

This gives us a second relationship between potential energy and force. In the case of the example we have been studying,

$$F_x = -\frac{dU}{dy} = -mg \tag{5}$$

as expected. If the derivative were difficult, which it is not, you could use Mathematica to find the force as follows:

In[4]:= **f = –∂$_y$ U[y]**

Graphing this function is unnecessary, but we will do it because we want to illustrate the use of similar techniques below.

In[5]:= **g2 = Plot[f, {y, 0, 10}, AxesLabel –> {"y", "F(y)"}];**

In some cases it is convenient to graph both the potential energy function and the related force:

In[6]:= **Show[GraphicsArray[{g1, g2}]];**

To summarize, associated with a conservative force is a potential energy function U. Taking the negative of the derivative of this function gives the force acting on the particle, and we may visualize the behavior of the particle by realizing that it will move to minimize U, keeping the total energy constant.

With these concepts it is possible to use the conservation of energy to find the velocity as a function of y. Then it is possible to integrate that equation to find y as a function of time. Finally, with two successive derivatives we can find the velocity and acceleration as a function of time. Thus, with no reference to Newton's second law, we can find exactly the same quantities as we can find with Newton's second law, albeit not as easily.

■ 6.3 One-Dimensional Single-Particle Systems

In[7]:= **Clear["Global`*"]**

In this section we want to look at the relationship between potential energy and force for several different systems, and we would like to visualize the behavior of the systems. Our forces, unlike the force of friction, will all be conservative so we can, in fact, define a potential energy function. We begin with the simple and move to the more complex.

Case I: $U(x) = k$

Suppose the potential energy is a constant k (possible zero). Then a graph of $U(x)$ is a straight, level line. Because Mathematica can only graph functions where the parameters have been previously defined, we will quite arbitrarily assume $k = 1$. Begin by defining the potential energy:

In[8]:= **U[x_] := k**
        k = 1;

Next, graph $U(x)$.

In[10]:= **g1 = Plot[U[x], {x, −2, 2}, AxesLabel −> {"x", "U(x)"}];**

Find the force:

In[11]:= **F = ∂_x U[x]**

and graph it

In[12]:= **g2 = Plot[F, {x, −2, 2}, AxesLabel −> {"x", "F(x)"}];**

6.3 One-Dimensional Single-Particle Systems

OK, so this one isn't very profound.

Finally, we make a graphics array of both graphs:

    In[13]:= **Show[GraphicsArray[{g1, g2}]];**

Now, with the potential energy graph in hand, how do we explain the behavior of a particle in this potential energy situation. This being a level surface, if we place a particle on it with an initial velocity of zero, it will have no tendency to move in the x-direction, and so it will stand still. The particle will always move with constant velocity.

Case II: $U(x) = kx$

We keep $k = 1$ and redefine U, the potential energy.

    In[14]:= **Clear["Global`*"]**
             U[x_] := k x
             k = 1;

Problem 1. Make a graph of the potential energy versus the position x, for positive x from 0 to 10.

Problem 2. By taking the derivative of the potential energy function, find the force acting on the particle.

Problem 3. Make a graph of this force for positive x from 0 to 10.

Problem 4. Make a graphics array of both of these graphs.

Problem 5. What would happen to a particle placed at some nonzero x?

Problem 6. This kind of potential energy system is very common. Identify it.

Case III: $U(x) = kx^2$

We begin with the definition and a plot.

    In[17]:= **Clear["Global`*"]**
             U[x_] = k x^2
             k = 1;

Now we are ready for some problems.

Problem 7. Make a graph of U for x from -4 to 4. Make sure you label the axes. This type of potential energy function is sometimes called a potential well. Why?

Problem 8. Find the force. Use:

In[20]:= $F = -\partial_x U[x]$

Problem 9. Graph the force and make a graphics array of both U and F.

Problem 10. Explain the behavior of the system in your own words.

Problem 11. Can you think of a system you have studied that has this potential energy function?

Also note that this is a familiar force function known as Hooke's Law (with $k = 1$).

Case IV: $U(x) = -\dfrac{k}{x}$

Problem 12. Use Mathematica to define the potential energy.

Problem 13. Make a graph of $U(x)$ versus x, $0 < x < 10$.

Problem 14. Find the force acting on the particle.

Problem 15. Make a graphics array showing both U and F.

Problem 16. Describe in words how the force varies with x; directly, inversely, directly as the square, inversely as the square, etc.?

Problem 17. Can you identify a system in nature that is similar to what we are describing?

Case V: $U(x) = \dfrac{1}{4x^4} - \dfrac{1}{2x^2}$

This is a much more complex system than what we have dealt with so far, and we shall do much of the work and the interpretation, leaving another difficult problem for you. Begin by defining the potential energy.

In[21]:= **Clear["Global`*"]**

$U[x\_] := +\dfrac{1}{4x^4} - \dfrac{1}{2x^2}$

Next, we graph it, using Mathematica's **PlotRange** option to select the features of the graph we expecially want to see.

In[23]:= **g1 =**
Plot[U[x], {x, 0, 5}, PlotRange -> {-.5, .5}, AxesLabel -> {"x", "U(x)"}];

Note that what we have is an asymmetric potential well with one side ($x \to 0$) having an infinite height and the other side ($x \to \infty$) a height of approximately 0.25. If it has a sufficiently small kinetic energy, a particle can be trapped in this potential well.

6.3 One-Dimensional Single-Particle Systems

The force can be found by taking the derivative. Thus

$$F[x\_] := -\partial_x\ U[x]$$

We plot the force and make a graphics array of the two graphs before attempting an interpretation.

In[25]:= g2 = Plot[F[x], {x, 0, 5}, PlotRange -> {-.5, .5},
         AxesLabel -> {"x", "F(x)"}];

In[26]:= Show[GraphicsArray[{g1, g2}]];

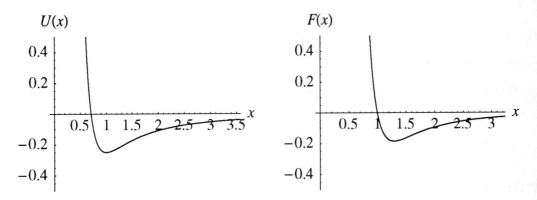

This potential energy function and force law is similar in *form* to the potential energy function of two atoms in a molecule with x representing the distance between the molecules. The potential energy function has a minimum somewhere between $x = 0.5$ and $x = 1.5$. The force, being the derivative of the potential energy function, must have a zero at this minimum, and we see this in the graph of F. We can also find this more exactly.

In[27]:= Solve[F == 0, x]

So indeed, the minimum is at $x = 1$. An atom placed at $x = 1$ will not move, it is in stable equilibrium. With some kinetic energy and a velocity directed toward the origin, the atom will move toward $x = 0$, but it cannot get there because the potential energy hill at $x = 0$ is infinitely high. If the atom at positive x moves toward the right with a small energy, it will lose speed, and bounce back; the atom is bound to the molecule. If the atom has a large kinetic energy, indeed if its total energy is positive, it will surmount the potential hill on the right and the molecule will disassociate.

Case VI: $U(x) = -\frac{1}{2} x^2 + \frac{1}{4} x^4$

Problem 18. Define the potential energy function in Mathematica.

Problem 19. Graph it on the interval $-2 \le x \le 2$. Discuss this function in terms of potential wells.

Problem 20. Find the force function and graph it on the same interval.

Problem 21. Make a graphics array of these two graphs.

Problem 22. Interpret the behavior of the system by thinking of a particle placed at some x with some initial kinetic energy. In particular, where is the particle in equilibrium? At which of these values of x is the particle in stable equilibrium? Unstable equilibrium? What will happen to the particle if it is placed at $x = 1$ with a kinetic energy that is too small to allow it to climb the hill to its left? What will happen to the particle if it is placed at $x = 1$ with a kinetic energy that allows it to go over the hill to its left? Can the particle ever escape from this system?

■ 6.4 Extensions to Two Dimensions

In[28]:= **Clear["Global`*"]**

In this section we simply extend the results of the previous sections to two dimensions. In other words, the potential energy function will be a function of x and y, and we might write it as $U(x, y)$. We can think of the graph of $U(x, y)$ as some topographical feature on the surface of the Earth, a hill for example, and our particles will tend to roll downhill. As a matter of fact, the systems we study all have radial symmetry and we can express the potential energy as $U(\mathbf{r})$ where $\mathbf{r}$ is a vector from the origin to a point in the x-y plane and $r^2 = x^2 + y^2$. Once again we proceed by cases.

| *Case I*: $U(r) = k, k = 1$

In our first case the potential energy function is a constant. Although this is a bit trivial, we graph it anyway to illustrate Mathematica's **Plot3D** command. We begin by defining the potential energy.

In[29]:= **Clear["Global`*"]**

$U[r\_] := k; \; r = \sqrt{x^2 + y^2}; \; k = 1;$

Then we graph it.

In[31]:= **Plot3D[U[r], {x, −2, 2}, {y, −2, 2}, Shading −> False, PlotPoints −> 20,
MeshStyle −> {Thickness[0.0001]}];**

We get exactly what we should have expected, a level surface. A particle placed at rest anywhere on the surface will not move. If it has some initial velocity, it will keep that velocity constant.

6.4 Extensions to Two Dimensions

Problem 23. What will happen to a particle if it is given some velocity toward the center of the well?

Problem 24. What will happen to a particle if it is given some velocity perpendicular to **r**?

Problem 25. Another kind of graph that is useful in looking at two-dimensional potential energy functions is the contour map. It gives results similar to the topographical maps published by the Geological Service. In this case the results are quite simple.

In[36]:= **ContourPlot[U[r], {x, −5, 5}, {y, −5, 5}, PlotPoints –> 70, Axes –> True, AxesLabel –> {"x", "y"}];**

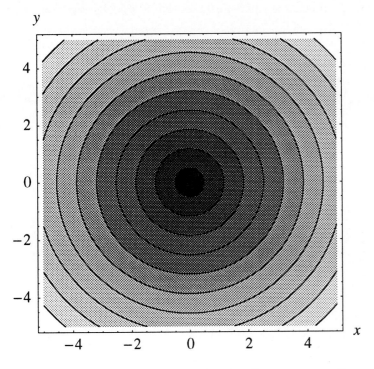

Interpret this graph. What do the circular lines represent?

In[37]:= **DensityPlot[U[r], {x, −5, 5}, {y, −5, 5}, Axes –> True, AxesLabel –> {"x", "y"}, Mesh –> False, PlotPoints –> 200];**

> Density plots look much better on the screen of a video monitor than printed, unless you have a very high resolution printer.

Case III: $U(r) = kr^2$

In[38]:= **Clear["Global`*"]**

Problem 26. Define the potential energy function and graph it with the **Plot3D** instruction.

Problem 27. Interpret the graph in terms of a particle placed at rest somewhere other than (0, 0).

Problem 28. Interpret the graph in terms of a particle given some velocity toward or away from the origin.

Problem 29. Interpret the graph in terms of a particle given some velocity perpendicular to **r**.

Problem 30. Make a contour map and density plot of the potential energy function.

| *Case IV*: $U(r) = -\dfrac{k}{r}$

We assign the exploration of this very important case as a project at the end of the chapter. This case corresponds to gravitational attraction as well as Couloumb attraction and repulsion, and although the problem is really three-dimensional, much can be learned by looking at its two-dimensional analogue.

| *Case V*: $U(r) = \dfrac{1}{4r^4} - \dfrac{1}{2r^2}$

Here we do everything in one fell swoop.

In[39]:= **Clear["Global`*"]; U[r_] :=** $\dfrac{1}{4r^4} - \dfrac{1}{2r^2}$**; r =** $\sqrt{x^2 + y^2}$ **;**
  Plot3D[U[r], {x, −4, 4},
   {y, −4, 4}, ColorFunction −> GrayLevel, PlotPoints −> 40,
   MeshStyle −> {Thickness[.0001]}, BoxRatios −> {1, 1, .75},
   Boxed −> False, Axes −> False, PlotRange −> {−.5, .3}];

6.4 Extensions to Two Dimensions

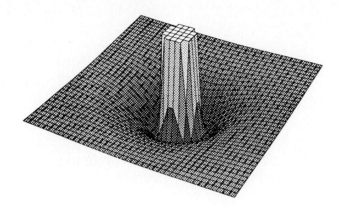

Here we have a potential energy peak in the middle of a potential well. Recall that this is representative of the potential energy function of two atoms separated by a distance r in certain diatomic molecules. An atom can orbit the other atom, perform radial oscillations, or do both. The atoms cannot assume a distance of zero, but with sufficient kinetic energy an atom can leave the potential well and the molecule will disassociate.

We can also make a contour map and a density plot.

> In[40]:= **ContourPlot[U[r], {x, −5, 5}, {y, −5, 5}, PlotPoints −> 100,
> PlotRange −> {−.20, 0}];**

> In[41]:= **DensityPlot[U[r], {x, −5, 5}, {y, −5, 5}, PlotPoints −> 200,
> Mesh −> False, ColorFunction −> Hue];**

Notice on the contour plot that the peak at the origin is represented by one single white spot.

Case VI: $U(r) = -\frac{1}{2} r^2 + \frac{1}{4} r^4$

Problem 31. Define the potential energy function.

Problem 32. Graph the potential energy function until you can see its essential features. We recommend a domain of {x, −1.5, 1.5}, {y, −1.5, 1.5}, and a plot range of {−.5, 0}. Also try a **ParametricPlot3D** plot.

Problem 33. Make a contour map and a density map of the potential energy function.

Problem 34. Write a short paragraph describing the behavior of a particle placed in this potential energy situation. Consider several possible initial velocities.

Mostly Mathematica

1. Make a **Plot3D** graph of the function

$$f(x, y) = \sin(xy) \tag{6}$$

for x in $[-2\pi, 2\pi]$ and y in $[0, 2\pi]$.

2. Find the minimum distance between the point $(1, 0)$ and the graph of the parabola $y = x^2$. Also, find the point on the parabola that has that distance. You will need the formula for the distance between two points.

3. Find the vector sum of **R**, a first-quadrant vector that has a magnitude of 25 and makes an angle of 34 degrees with respect to the $+x$-axis, **S**, a second-quadrant vector with magnitude 47 and that makes an angle of 63 degrees with respect to the $-x$-axis; and **T**, a third-quadrant vector that makes an angle of 45 degrees with respect to the $-y$-axis and has a magnitude of 33.

Explorations

1. Study the system with the potential energy function $U(r) = -k/r$. Use Mathematica to define and graph the potential energy function. The gravitational potential energy of a mass m in the vicinity of a mass M is $-GMm/r$, and therefore this potential energy function is of great interest. In particular, discuss the possibility of orbits with this potential energy function. Note that in Chapter 4 we constructed orbits for this function. Finally, we should mention that in nature **r** is actually a three-dimensional vector and the potential energy function becomes rather difficult to visualize. Try it. If you are interested, there are, in fact, Mathematica packages to help visualize these situations. The package is **Graphics `PlotField3D`**.

2. The potential energy of a mass m in the vicinty of a binary star system, each star having a mass M, is

$$U(r) = -\frac{GMm}{r_1} - \frac{GMm}{r_2} \tag{7}$$

where r_1 is the distance of m from one star and r_2 is the distance of m from the other star.

To graph this potential energy function, first assume $GMm = 1$. Next, draw an x-y coordinate system, and place one star at the point $(-1, 0)$ and the other at $(1, 0)$, and place the mass m at some arbitrary point (x, y). Sketch the distances r_1 and r_2. Recall from analytic geometry that the distance from (x, y) to $(1, 0)$ is

6.4 Extensions to Two Dimensions

$$r_1 = \sqrt{(x-1)^2 + y^2} \,. \tag{8}$$

Write a corresponding expression for r_2. Proceed to write an expression for the potential energy $U(x, y)$. Finally, graph this expression to obtain a figure such as the one that begins the chapter. Also make a contour map and a density map of this potential energy function and discuss the behavior of a particle in this system, including the possibility of orbits.

3. Find a one-dimensional potential energy function that has a minimum at $x = 0$, maxima at $x = -1$ and $x = 1$, but no other local maxima or minima. Graph it, and discuss the behavior of a particle in this system.

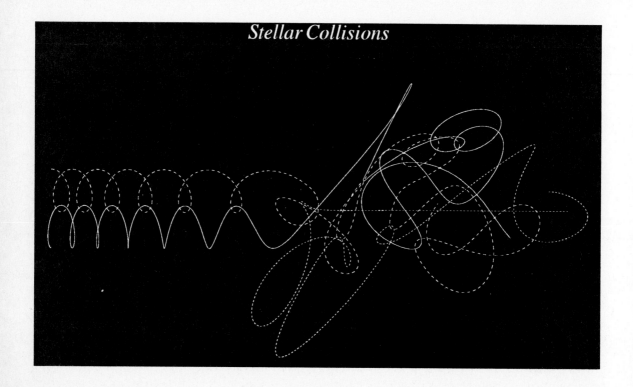

Stellar Collisions

CHAPTER 7 Newton's Second Law Is a Differential Equation

■ 7.1 Introduction

I must confess the title to this chapter is not mine. The title belongs to a paper given by a professor at a meeting of physics professors, someone whose name is unfortunately now forgotten. He might be happy to know that some of the spirit of his paper has not been lost.

Most of us think of Newton's second law as follows:

$$\mathbf{F} = m\mathbf{a} \qquad (1)$$

Part of the rationale for writing the law in this way is to avoid intimidating students like yourself with the fact that this is a differential equation, an equation with one or more derivatives in it. The $\mathbf{F} = m\mathbf{a}$ form of the law is deceptively simple in two ways. First, the force $\mathbf{F}$ is, in reality, the *sum* of all of the external forces acting on the object. Secondly, the acceleration **a** is really the second derivative of the position vector **r** of the center-of-mass of the object. Thus, a more precise way to write Newton's second law is

$$\sum \mathbf{F}_{external} = m\mathbf{r}''(t) \qquad (2)$$

Now you see that Newton's second law is a differential equation with the derivative

$$\mathbf{r}''(t) = \frac{d^2\mathbf{r}}{dt^2} \qquad (3)$$

in it. In fact it is a second-order differential equation. Because it may be another semester or more before you take a course in differential equations, the physics problems you encounter are generally those in which **a**(t) can be integrated directly to give **v**(t), and it may be integrated directly to give **r**(t). However, there are much more complex situations and we shall look at just a few in this book.

As a matter of fact, differential equations are ubiquitous in physics. The most important laws of physics are in the form of differential equations. Newton's second law of motion is the fundamental postulate of classical physics, and the fundamental postulate of quantum mechanics is also a differential equation. Because the mathematical expertise of students taking a first course in calculus-based physics does not include solving differential equations, they are for the most part excluded from this course. However, Mathematica has commands that solve differential equations and, as long as you are extremely careful, you may experiment with this command to solve various problems in physics. You must be careful because your intuition about the solutions to differential equations is rather dramatically underdeveloped at this point, and you might not only feel but also be helpless when commands do not work as expected or you get error messages for your efforts. Moreover, the experience you get in this chapter will fall far short of making you an expert.

■ 7.2 Starting Simple: A Ball Thrown Upward

It will be very useful in what follows to use the prime notation for derivatives and to write Newton's second law as two first-order differential equations rather than one second-order differential equation. Thus, Newton's second law becomes

$$\sum F_{ext} = mv'(t), \quad x'(t) = v(t) \qquad (4)$$

where we have dropped the vector notation because we wish to work a few one-dimensional problems. Writing the differential equation this way allows us to find both the position and the velocity as functions of time. To begin, it would also be a good idea to perform a **Clear** and to determine the syntax of the command, **DSolve**, which solves differential equations, like this:

In[1]:= **Clear["@"]**

In[2]:= **?? DSolve**

Problem 1. To illustrate the use of **DSolve**, we consider the motion of a 1-kg point mass thrown from the ground upward into the air with an initial velocity of 50 m/s. Establish a coordinate system with its origin at ground level and the *x*-axis pointing upward, and write the kinematics equations for motion with constant acceleration that govern the motion of this object. You probably learned these early in your physics course.

Now we will obtain these equations with Mathematica's **DSolve** command. Neglecting air resistance, Newton's second law becomes

$$-mg = mv'(t) \text{ or } x'(t) = v(t). \qquad (5)$$

We assume the same coordinate system as yours. For initial conditions we assume $x(0) = 0$ and $v(0) = 50$. Noting that the mass cancels in Newton's second law, we enter the equations into the **DSolve** command as follows:

In[3]:= **solution = DSolve[{v'[t] == −g, v[0] == 50}, v[t], t]**

> Always use the double-equals signs, ==, in **DSolve**. Differential equations are conditional equations, not assignments. Also note that **solution** is just a name assigned to the answer; we could just as well use another name, **sol**, **newton**, or **jack** for examples.

Notice that we have only obtained a solution for $v(t)$. It is much more efficient to find solutions for *v* and *x* simultaneously, as follows:

In[4]:= **solution =
DSolve[{v'[t] == −g, v[0] == 50, x'[t] == v[t], x[0] == 0}, {v[t], x[t]}, t]**

> To the extent possible, use symbols (*g,* for example) rather than actual values (9.8 for example) in the **DSolve** command. Also, avoid decimal numbers whenever possible. Assign parameter values after the differential equations are solved.

Do these solutions agree with your textbook solutions?

Next, we assign *g* a value and plot the solutions:

In[5]:= **g = 9.8
Plot[v[t] /. solution, {t, 0, 10}];
Plot[x[t] /. solution, {t, 0, 10}];**

How long is the object in the air? This command produces the answer:

7.2 Starting Simple: A Ball Thrown Upward

In[7]:= **Solve[x[t] == 0 /. solution, t]**

This instruction means: Find the time t when $x = 0$, given the solution to the differential equation.

How high does it go? This pair of commands produce the answer.

In[8]:= **Solve[v[t] == 0 /. solution, t]**

In[9]:= **x[5.10204]**

Why use the time at which the velocity is zero?

Problem 2. An object is thown upward from the edge of a cliff 20 m high. The initial speed is 40 m/s. Use the **DSolve** command after a **Clear** to find x and v as functions of time. How high above the ground will it go? How long does it stay in the air? With what speed does it hit the ground? Answer these questions using Mathematica and the techniques and commands you have already seen in this chapter.

Problem 3. To give you some additional practice with **DSolve**, consider this first-order differential equation:

$$\frac{dN}{dt} = -kN(t) \text{ or } N'(t) = -kN(t) \tag{6}$$

Equation (6) says the rate of change of a quantity N is proportional to the quantity itself. Money invested in a continuously compounded savings account behaves this way (plus sign), while the number of radioactive atoms in a sample of atoms may also behave in this way (minus sign). You will see this equation again and again in physics, biology, and engineering. Let's **DSolve** it. Let $N(0) = A$, the original amount of stuff (money, atoms, rabbits) First we **Clear**,

In[10]:= **Clear["Global`*"]**

then we solve the differential equation (with the + sign), as follows:

In[11]:= **solution = DSolve[{N'[t] == k N[t], N[0] == a}, N[t], t]**

Graph the solution, $N(t)$, assuming $a = 100$ and $k = 1$. A solution such as this is sometimes called an explosion. Why?

Problem 4. Solve the same differential equation, but with the minus sign. Graph your solution and describe in your own words what happens to the quantity $N(t)$ as a function of time.

Problem 5. We consider the coffee-cooling problem. The rate-of-change of the temperature of coffee in a cup is given by

$$T'(t) = -k(T(t) - T_e) \tag{7}$$

where $T(t)$ is the temperature of the coffee as a function of time, T_e is the temperature of the environment of the cup, and k is a constant which is determined primairly by the insulating properties of the cup. Assume $T_e = 25$ degrees, $T = 98$ degrees at $t = 0$, and $k = 0.1$. Plot the temperature as a function of time. Interpret the graph in your own words.

Problem 6. To give you some additional practice with **DSolve**, here is another problem. Newton's second law applied to a mass m on a spring with force constant k is

$$-kx(t) = mx''(t) \tag{8}$$

Begin by writing this differential equation as two first-order differential equations. For simplicity assume $m = k = 1$. For initial conditions choose $x = 1$ and $v = 0$. Put this information into the **DSolve** command, give the instruction a label such as **solution**, and execute the command.

Problem 7. Graph both **x[t]** and **v[t]** for **t = 0** to **t = 4π**.

Problem 8. Use **ParametricPlot** as follows to graph **v[t]** versus **x[t]** as follows:

In[12]:= **ParametricPlot[{x[t], v[t]} /. solution, {t, 0, 2π}];**

Interpret the results of **ParametricPlot**, called a phase portrait.

Problem 9. Try a different set of initial conditions, namely $x = 0$ and $v = 1$ at $t = 0$. How does this solution compare with the previous solution?

■ 7.3 Objects Falling in a Resistive Medium

In Chapter 3, we used a numerical analysis technique to study objects falling in air. Here we use **DSolve** to obtain similar results for an object that falls in a resistive medium. We need to consider a couple of cases depending on whether the resistive force is proportional to the speed or the square of the speed.

7.3.1 Resistive Force Proportional to v^2

An object falls under the influence of gravity and the force of air resistance which we imagine to be proportional to the square of the speed. If we take positive as downward, then the equation of motion is

$$mv'(t) = mg - kv^2 \tag{9}$$

7.3 Objects Falling in a Resistive Medium

With a little tricky algebra, which you should try to reproduce, we may rewrite this equation as follows:

$$g\left(1 - \frac{v^2}{v_T^2}\right) = v'(t) \tag{10}$$

where v_T is the terminal velocity

$$v_T = \sqrt{\frac{mg}{k}} \tag{11}$$

We will also assume initial conditions $x = v = 0$ at $t = 0$; in other words, we are simply dropping the object. Now we are ready for **DSolve**. Using **c** for v_T to avoid subscripts in Mathematica commands, we write

In[13]:= **Clear["@"]**
solution =

$$\textbf{DSolve}\left[\{\textbf{v'[t]} == \textbf{g}\left(1 - \frac{\textbf{v[t]}^2}{\textbf{c}^2}\right), \textbf{x'[t]} == \textbf{v[t]}, \textbf{v[0]} == 0, \textbf{x[0]} == 0\},\right.$$
$$\left.\{\textbf{v[t], x[t]}\}, \textbf{t}\right]$$

> The error message is far from indicating anything fatal. However, be aware that whenever inverse functions are used to solve equations, not all solutions are found. A simple example is to solve **Sin[x]** == $\frac{1}{2}$. Only one solution is found, when there are an infinite number of solutions.

Here we see that the solution for $v(t)$ involves the hyperbolic tangent and the solution for $x(t)$ involves the hyperbolic cosine, and moreover, two of the three solutions involve complex numbers. We can select the first two (real) elements of the solution with this command.

In[15]:= **realsolution = solution[[1]]**

> The notation **solution[[1]]** selects the first element in the set of three solutions of **DSolve**.

We graph the solutions to see if they agree with what we learned in Chapter 3, choosing more or less arbitrarily a terminal velocity of 10 m/s.

In[16]:= **Plot[v[t] /. realsolution /. {g -> 9.8, c -> 10}, {t, 0, 3}, PlotRange -> All, AxesLabel -> {"t", "v(t)"}];**

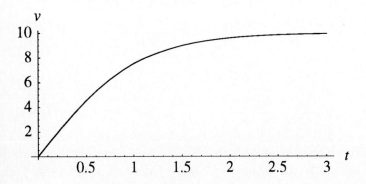

In[17]:= **Plot[x[t] /. realsolution /. {g -> 9.8, c -> 10}, {t, 0, 3}, PlotRange -> All, AxesLabel -> {"t", "x(t)"}];**

Problem 10. How do these curves for $x(t)$, $v(t)$, and $a(t)$ differ from those without air resistance? Graph both on the same coordinate system either by plotting some points on the curves given above or by using Mathematica to superimpose both graphs.

We can find the acceleration by taking the derivative with respect to time, as follows:

In[18]:= **a[t_] = ∂_t (v[t] /. realsolution /. {g -> 9.8, c -> 10})**

Next we plot the acceleration as a function of time.

In[19]:= **Plot[a[t], {t, 0, 3}, AxesLabel -> {"t", "a(t)"}];**

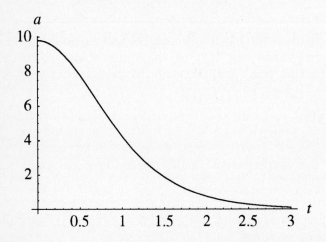

Problem 11. On the preceding graph of the acceleration plotted by Mathematica, sketch a graph of the acceleration that would occur if there were no air resistance.

7.3.2 Resistive Force Proportional to v

In[20]:= **Clear["@"]**

You get to do this problem by working your way through these exercises.

Problem 12. Similar to the process above, use Newton's second law to write a differential equation for a falling object assuming that the force of the air resistance is proportional to the speed rather than the speed squared.

Problem 13. Express the equation in terms of the terminal velocity,

$$v_T = \frac{mg}{k} \tag{12}$$

Problem 14. Use the same initial conditions, terminal velocity, and acceleration of gravity as we did above and **DSolve** the differential equation. Be sure to first use a **Clear** command.

Problem 15. Plot the solutions for x and v as functions of time.

Problem 16. Find and plot the acceleration.

Problem 17. This is not an easy question, but what are the main differences between a linear and a quadratic dependence of the resistive force on the speed, given the same terminal velocity? Which one gives the best fit to the shuttlecock data given in Chapter 3?

■ 7.4 A Problem in Two Dimensions

In[21]:= **Clear["@"]**

Mathematica's **DSolve** can also handle differential equations in several dimensions. Perhaps the simplest example is projectile motion without air resistance. Here are the differential equations we obtain from Newton's second law.

$$mv'_x(t) = 0, \; mv'_y(t) = -g, \; x'(t) = v_x(t), \; y'(t) = v_y(t) \tag{13}$$

Let us place our coordinate system at the point where the projectile starts on its trip through the sky, namely at ground level. For initial conditions we have $x = y = 0$, and to get the projectile to fly, we choose to send it off at $t = 0$ with an x-component of velocity v_{x_0} and a y-component v_{y_0}. The physics is now complete, and we turn to the mathematics.

In[22]:= **projectile = DSolve[{vx'[t] == 0, vy'[t] == −g, x'[t] == vx[t],
     y'[t] == vy[t], x[0] == 0, y[0] == 0, vx[0] == vx0, vy[0] == vy0},
     {vx[t], x[t], vy[t], y[t]}, t]**

You should recognize the various parts of the solution from the kinematics problems you did earlier in the semester, so the answers are not too exciting. Let's do a few things with our results anyway, to illustrate how to deal with them. How do we plot the results? We need to specify the initial conditions and parameters first. Let's do that all at once with the following expression:

In[23]:= **rules = {vx0 −> 30, vy0 −> 40, g −> 9.8};**

Now we find when the projectile hits the ground.

In[24]:= **Solve[(y[t] /. projectile /. rules) == 0, t]**

Next we plot the trajectory:

In[25]:= **ParametricPlot[{x[t], y[t]} /. projectile /. rules, {t, 0, 8.16},
     AxesLabel −> {"x(t)", "y(t)"}];**

> Ignore the error message or eliminate it by adding a **Compiled −> False** option to **ParametricPlot**.

Let's see if the total energy (per unit mass) is conserved during the flight. Assume the mass of the projectile is 1 kg.

In[26]:= **Plot[(1/2 (vx[t]^2 + vy[t]^2) + g y[t]) /. projectile /. rules, {t, 0, 8},
     PlotRange −> {0, 1300}];**

Problem 18. Find *exactly* how high the baseball goes, how far it goes, and how long it is in the air, given the same initial conditions.

Problem 19. Repeat Problem 18 with the initial conditions are $v_{x_o} = 40$ and $v_{y_o} = 30$.

Problem 20. A model of running due to Keller (1973) yielded this differential equation for the velocity of a runner in a sprint:

$$f(t) = v'(t) + \frac{v(t)}{\tau} \qquad (14)$$

where $f(t)$ is the force per unit mass supplied by the runner, and τ is a parameter that is a property of the runner. For sprints $f(t)$ is (approximately) a constant. Use **DSolve** to find an expression for $v(t)$ and $x(t)$. It should be clear what initial conditions are appropriate. Does increasing τ increase or decrease the speed of a runner? For more details, including data, see Wagner (1998).

Choose some value for f and τ (try 8 and 1.5, respectively) and graph the results.

In[27]:= **Plot[v[t] /. (solution /. {f –> 8, τ –> 1.5}), {t, 0, 10}];**

Problem 21. A refinement of the previous model has a nonconstant force

$$f(t) = F - ct \qquad (15)$$

where F and c are constants. In this case the force decreases with time as the runner tires. Add this complication to the differential equation and solve it. For some trial values of the constants choose $F = 8.7$, $\tau = 1.42$, and $c = 0.066$: according to Wagner (1998) these are for a run of Ben Johnson at a world championship. Plot $v(t)$ and compare it to the previous result. What is the main difference in the velocity curve between the situation where the force is constant and where it decreases with time? (Perhaps you could apply this theory to horse racing and actually use physics to make money!)

■ 7.5 Numerical Solutions of Differential Equations

That not all differential equations have analytic solutions sometimes comes as a surprise to students. In other words, it might be *impossible* to find a function that satisfies the differential equation. What do you do in that case? You either use the numerical integration techniques we introduced in the early chapters of this book or some other integration technique and find a numerical solution. Not surprisingly, Mathematica has a command, **NDSolve**, which finds numerical solutions to differential equations. It is a sophisticated, versatile, and powerful command that is implemented by approximately 500 pages of code in the computer language known as C.

In the first several chapters of this book we also solved differential equations with a technique known as Euler's method or the Euler-Cromer method. Why didn't we just use **NDSolve**? Because it is important to understand the foundation upon which numerical solutions are based, which you should now understand, and because it is important to understand the various quantities that are involved, step-size h for example, as well as the problems you will encounter. Only then can you begin to understand and appreciate the capabilities as well as the options of **NDSolve**.

We begin looking at **NDSolve** with a simple problem that we studied earlier and that has an analytic solution. This problem can be solved with **DSolve** and compared with the solution provided by **NDSolve** A cup filled with hot coffee at an initial temperature of 98 °C is in surroundings at 25 °C. The differential equation governing the temperature of the coffee is

$$T'(t) = -k(T(t) - T_e) \qquad (16)$$

What this expression says is that the rate of change of the temperature of the coffee is proportional to the temperature difference between the coffee and the environment. This is true because the rate of energy flow across a boundary is proportional to the temperature difference across the boundary. We will let T in this equation be represented by **temp** in Mathematica. We have made an arbitrary choice of 0.1 for k. Notice that we begin with a **Clear**, and then proceed to **NDSolve**. Pay particular attention to how the differential equation in its more or less standard form gets translated into Mathematica. Also note how the initial condition is stated and that some interval of time must be specified. Our solution will cover only that time interval.

Finally, take note of the output of this command, which is an *interpolating function*. The traditional form of a numerical solution to a differential equations is a table of numerical points $(t, T(t))$ such as we found in Chapters 2 and 3. Mathematica makes the choice of replacing the table with a function whose graph fits the points in the table. This allows you to find values of the temperature between the entries in the table by interpolation.

In[28]:= **Clear["Global`*"]**
solution = **NDSolve[{T'[t] == −.15 (T[t] − 25), T[0] == 98}, T, {t, 0, 30}]**

In[30]:= **Plot[T[t] /. solution, {t, 0, 30}, AxesLabel −> {"t", "T(t)"},
    PlotRange −> {0, 100}];**

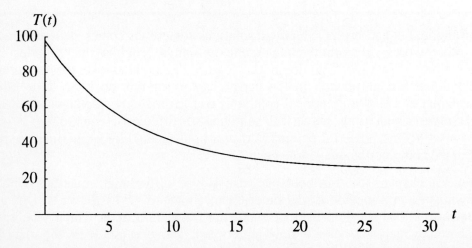

Notice how the graph approaches the environmental temperature asymptotically.

Problem 22. Use **NDSolve** to find a numerical solution of the differential equation

$$N'(t) = N(t) \qquad (17)$$

for the interval $0 \leq t \leq 4$ with initial condition $N(0) = 100$. Graph the solution. Use graphs to compare the numerical solution to the analytical solution provided by **DSolve**. Also use the **Table** command to compare the two solutions.

7.5 Numerical Solutions of Differential Equations

Now that we have one example of using **NDSolve**, we turn our attention to a two-dimensional problem, namely projectile motion with air resistance. We have considered this problem in Chapter 4, and you might wish to refer to that material to see what we did. Here we will show how to handle the same problem with Mathematica's **NDSolve** command. In Chapter 4 we showed that the equations that govern the motion are

$$v'_x(t) = -kvv_x(t) \text{ and } v'_y(t) = -g - kvv_y \qquad (18)$$

with

$$k = \frac{1}{2} C\rho_a \frac{A}{m} \text{ and } v = \sqrt{v_x^2 + v_y^2} \qquad (19)$$

where C is the drag coefficient, A is the cross-sectional area of the object perpendicular to the direction of motion, m is the mass of the object, ρ_a is the density of air, and v is the magnitude of the velocity. Here is how these equations are translated into Mathematica's **NDSolve** command. We begin by clearing.

In[32]:= **Clear["Global`*"]**

Next, we give all of the parameters numerical values, in this case those of a baseball traveling in air near the surface of the Earth.

In[33]:= **k = 1/2 c ρ $\frac{A}{m}$ /. {c -> 0.46, ρ -> 1.2, A -> π 0.0367^2, m -> 0.145}**

g = 9.8

Now we define the magnitude of the velocity.

In[35]:= **v[t_] := $\sqrt{vx[t]^2 + vx[t]^2}$**

Here is the way the problem is finally solved.

In[36]:=
    solution = NDSolve[{vx'[t] == -k v[t] vx[t], vy'[t] == -g - k v[t] vy[t],
        x'[t] == vx[t], y'[t] == vy[t], vx[0] == 30,
        vy[0] == 40, x[0] == 0, y[0] == 0}, {x, y, vx, vy}, {t, 0, 10}]

Next, we graph the solution.

In[37]:= **ParametricPlot[{x[t], y[t]} /. solution, {t, 0, 6.1},**
    AxesLabel -> {"x(t)", "y(t)"}, Compiled -> False];

Problem 23. Solve some of the baseball problems in Section 4.6.

We will look at some additional problems involving **NDSolve** in the next chapter. Although we have worked with some powerful commands in this chapter, it is important to keep a sense of humility about our knowledge of differential equations. First of all, these commands are of absolutely no help if you cannot find the differential equation that applies to the particular physical situation. Second, differential equations is a complex field and it requires years of experience before one can be identified as an expert. On the other hand, there is no reason why you should not enjoy the power of **DSolve** and **NDSolve** as you explore physics.

Mostly Mathematica

1. Solve and graph the solution to the differential equation

$$x''(t) = -x(t) \tag{20}$$

Assume $x(0) = 0$ and $x'(0) = 1$. Find a numerical solution of the same equation.

2. Write the previous differential equation as two first-order differential equations, then solve for $x(t)$ and $v(t)$ ($v(t) = x'(t)$). Assume the same initial conditions. Make a **ParametricPlot** with $x(t)$ along the x-axis and $v(t)$ along the y-axis.

3. Verify by taking derivatives and making substitutions that

$$y(t) = \frac{k^2}{g} \ln\left(\cosh\left(\frac{gt}{k}\right)\right) \tag{21}$$

is a solution to the differential equation

$$y''(t) = g\left(1 - \left(\frac{y'(t)}{k}\right)^2\right) \tag{22}$$

Begin by writing $y(t)$ as a function in Mathematica. Then find each side of Equation (22) individually and see if they are the same. Remember, in Mathematica ln is **Log**, cosh is **Cosh**, and all functions must have their arguments in square brackets. Are tanh and sech related by a theorem?

Explorations

1. Take the coffee-cooling problem, for example, and solve it analytically with **DSolve** and numerically with **NDSolve**. Make tables of data for both solutions and compare.

2. Compare the solutions to the two-dimensional motion problems of the baseball in air obtained from **NDSolve** and our own Euler-Cromer method in Section 4.6. Do this by making tables of the results.

3. Solve the rocket-to-the-Moon problem with **NDSolve** rather than the Euler-Cromer method. Refer to Section 4.8.

4. Look ahead and try to solve the differential equations associated with *RC* circuits. Charging a capacitor through a resistor leads to this differential equation:

$$V - Rq'(t) - \frac{1}{C} q(t) = 0 \qquad (23)$$

When discharging the capacitor, the differential equation is:

$$Rq'(t) + \frac{1}{C} q(t) = 0 \qquad (24)$$

5. Solve the problem of a satellite orbiting a massive object.

6. Model the motion of a particle under the influence of a force described by

$$F = \frac{1}{x^5} - \frac{1}{x^3} \tag{25}$$

Try various initial conditions.

7. Suppose a person or a car accelerates with constant power P. In that case, the differential equation governing the motion is

$$P = mv \frac{dv}{dt} \tag{26}$$

Remember, P is a constant. Solve this differential equation with **DSolve** and graph the results. Does this model really apply to an Olympic sprinter or an accelerating Miata?

References

Keller, J. *Phys. Today* 26 (1973): 43.

Wagner, G. The 100-meter dash: Theory and experiment, *The Physics Teacher* 36 (March 1998): 144.

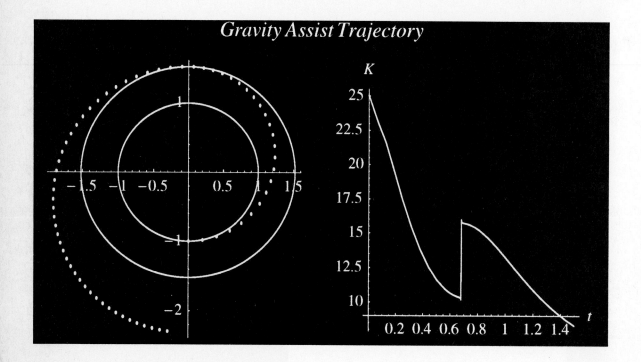

CHAPTER 8 Topics in Linear Momentum and Gravitation

■ 8.1 Introduction

In this chapter we will begin by modeling a simple elastic collision between two particles. This will give us more experience with **NDSolve** in a context with which we are familiar. Then we will go on to use this same command in a number of more complex situations involving gravity and the solar system including gravity-assist orbits made famous by the *Voyager* spaceships and, of more dubious distinction, the film *Armageddon* in which the space shuttles use the Moon as the slingshot agent to catch up with an asteriod hurtling toward Earth. Finally, we will look at a problem from biology. You may wish to refer to our use of **NDSolve** in Chapter 7, and you may wish to refer to your textbook while working your way through these problems. Many texts have a separate section on orbital motion and planetary motion.

8.2 Linear Momentum

In[1]:= **Clear["Global`*"]**

We intend to study the conservation of momentum by creating some collisions. Of course, in the lab you can just pick up an object, pitch it at another object and watch what happens. Forces between the objects come into play automatically when they hit. In the computational lab we must invent a force and use the laws of physics, specifically Newton's second law to create this collision.

We will assume a force law between two objects, molecules perhaps, of the form

$$F(x) = \frac{1}{x^7} - \frac{1}{x^5}, \qquad (1)$$

where x is the distance between the particles. We define the force and graph it.

In[2]:= **F[x_] := $\frac{1}{x^7} - \frac{1}{x^5}$**
Plot[F[x], {x, 0, 4}, PlotRange -> {-.5, 1}, AxesLabel -> {"x", "F(x)"}];

This command graphs the force as a function of distance. Note that when x is small the force becomes repulsive, while for $x > 1$ it is attractive. We can also find

In[4]:= **Limit[F[x], x -> Infinity]**

and observe that for large enough separation the force is essentially zero.

Here is the plan. We will put one particle, mass = **m1**, at some initial position, **x1[0]**, and give it some initial velocity **v1[0]**. We will put another particle, mass = **m2**, at some other initial position, **x2[0]**, and also give it some initial velocity, **v2[0]**, preferably chosen so that the particles collide. Here are some initial conditions with which to try our experiment. We can change them later if we like.

In[5]:= **initcond = {x1[0] == 0, x2[0] == 10, v1[0] == 2, v2[0] == -2}**

Problem 1. What will happen to the particles with these initial conditions?

The distance between the masses must always be a positive quantity, hence

In[6]:= **x[t] = Abs[x1[t] - x2[t]]**

Here is Newton's second law for mass **m1**. We assume mass **m2** is off to the right. That means the repulsive force on **m1**, call it **F1**, should be to the left, hence negative when the distance between the masses is less than one. Thus,

In[7]:= **F1[x_] = –F[x]**

We write Newton's second law for mass **m1** as a first-order differential equation.

In[8]:= **eqn1 = m1 v1'[t] == F1[x[t]]**

Using Newton's third law we may write

In[9]:= **F2[x_] = –F1[x]**

to find the force on mass **m2**. Thus

In[10]:= **eqn2 = m2 v2'[t] == F2[x[t]]**

Since we would like to know the positions as well as the velocities as a function of time, we also write first-order differential equations for **x1[t]** and **x2[t]**.

In[11]:= **eqn3 = x1'[t] == v1[t]**
          eqn4 = x2'[t] == v2[t]

To begin our experiments, let

In[13]:= **m1 = 1; m2 = 1**

Finally, we collect all our equations in one set and put them into the **NDSolve** command.

In[14]:= **solution = Flatten[NDSolve[Union[{eqn1, eqn2, eqn3, eqn4}, initcond],**
          {x1, x2, v1, v2}, {t, 0, 5}]]

> The use of **Flatten** in this command may seem mysterious. In fact, most, but not all, of the results of this section can be obtained without **Flatten**. Mathematica gives its solutions to equations, including differential equations, as sets of sets, and all that **Flatten** does in this situation is get rid of one set of set brackets. If we do not **Flatten** at this point, then when we attempt to integrate, below, we will fail to get an answer.

The output of **NDSolve** is an interpolating function that we proceed to graph as follows:

In[15]:= **Plot[{x1[t] /. solution, x2[t] /. solution}, {t, 0, 5}, AxesLabel –> {"t", "x"}];**

8.2 Linear Momentum

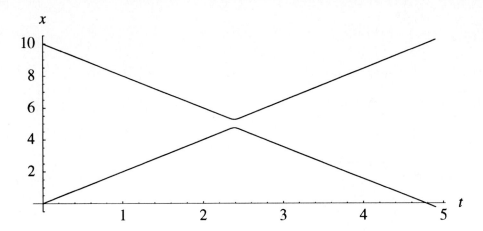

Problem 2. Interpret this graph. Which curve corresponds to particle 1? Which corresponds to particle 2? What happens to particle 1? Interpret the slopes of the two curves.

Problem 3. Make a prediction about how the velocities of the two particles will change. Use the graph below (or a copy) to sketch the velocity of each.

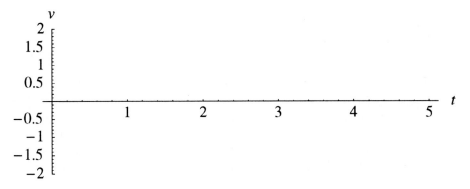

Problem 4. Next, use Mathematica and the interpolating functions of the solution to graph the velocities. How did you do on your prediction? What was right and what was wrong?

Problem 5. Another interesting graph to see is a graph of the force versus time. Here we graph the force of **m2** on **m1**. Interpret it.

$$\text{In[16]:= Plot}\left[\frac{1}{(x1[t] - x2[t])^7} - \frac{1}{(x1[t] - x2[t])^5}\text{ /. solution, \{t, 1, 4\},}\right.$$
$$\left.\text{PlotRange -> All, AxesLabel -> \{"F", "t"\}}\right];$$

Problem 6. Recall from Chapter 6 that the potential energy of this system is

In[17]:= $U[x\_] = -\int_{\infty}^{x} F[x]\,dx$

where we have chosen our zero for potential energy as $x \to \infty$. Therefore

In[18]:= **pe = U[x[t]] /. solution**

Plot the potential energy, **pe**, for t from zero to five.

In[19]:= **Plot[pe, {t, 0, 5}, PlotRange −> All, AxesLabel −> {"t", "U"}];**

Problem 7. What is the kinetic energy, K, of the system as a function of time? Predict the graph of K on this coordinate system (or a copy):

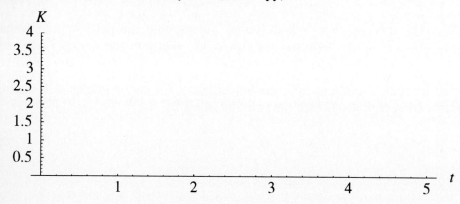

Then, use Mathematica to make a graph of kinetic energy versus time. Comment on your prediction. If you mentally add the graphs of potential energy and kinetic energy (as a function of time), does it appear that energy is conserved during the collision? Use Mathematica to graph $U + K$.

Clearly we have handled one of the simpler cases. Let's look at one additional case in which particle 2 is initially at rest and has a mass of $\frac{1}{10}$ the mass of **m1**. We group all of the instructions in one cell so that we can execute them all in one step.

In[20]:= **Clear["Global`*"]**

$F[x\_] := \dfrac{1}{x^7} - \dfrac{1}{x^5}$;

initcond = {x1[0] == 0, x2[0] == 10, v1[0] == 2, v2[0] == 0};
x[t] = Abs[x1[t] − x2[t]];
F1[x_] = −F[x];
eqn1 = m1 v1'[t] == F1[x[t]];
F2[x_] = −F1[x];
eqn2 = m2 v2'[t] == F2[x[t]];

8.2 Linear Momentum

```
eqn3 = x1'[t] == v1[t];
eqn4 = x2'[t] == v2[t];
m1 = 1; m2 = 1/10;
solution = Flatten[NDSolve[
        Union[{eqn1, eqn2, eqn3, eqn4}, initcond], {x1, x2, v1, v2}, {t, 0, 10}]];
Plot[{x1[t] /. solution, x2[t] /. solution}, {t, 0, 10}, AxesLabel -> {"t", "x"}];
```

The graph clearly shows that the large mass **m1** slows only slightly after the collison, while the small particle has a high speed after the collision. That probably does not surprise us. Here is a related question. If a monarch butterfly heads up a highway and is hit by an 18-wheeler, which receives the larger impulse, the butterfly or the 18-wheeler?

The situation we have just modeled is not quite as extreme as the butterfly and the 18-wheeler, but it can help us get an answer. Recall how impulse is defined:

$$\text{impulse} = \int_{t_1}^{t_2} F(t)\,dt, \qquad (2)$$

where $F(t)$ is the force acting on the particle. We can find the variation of the force on **m1** with time as follows:

In[22]:= **P1 = F1[x[t] /. solution]**

and, therefore, the impulse on **m1**, the larger mass, is

In[23]:= **NIntegrate[P1, {t, 0, 8}]**

where we have integrated over most of the collision time.

> We must do a numerical integration because we do not have the exact functional dependence of **F1** with time, instead we have an interpolating function, which essentially gives us a table of values.

Next, we find the impulse on **m2**.

In[24]:= **P2 = F2[x[t] /. solution]**

In[25]:= **NIntegrate[P2, {t, 0, 8}]**

and we see that the impulses on the two particles are identical. Given Newton's third law, this result should never suprise us, nor should the answer to the butterfly problem.

Problem 8. Predict what happens when a small particle collides with a large one. Then model this situation using the tools we have developed.

Problem 9. Predict what happens when two equal masses having equal but opposite velocities collide. Model this situation using the tools we have developed.

8.3 Motion Under the Influence of Gravity

During the remainder of this chapter we shall look at a variety of problems involving gravity and, more specifically, motion of objects in the solar system. The power of **NDSolve** makes a large number of problems tractable. We will begin with the simplest system, an object in orbit around the Sun, which we consider so massive compared to the orbiting object that we may neglect its own motion. The force on the orbiting object is, of course, given by Newton's universal law of gravitation:

$$F = \frac{GMm}{r^2} \tag{3}$$

8.3.1 Halley's Comet

In[26]:= **Clear["Global`*"]**

To begin, draw a coordinate system, label the x- and y-axes, and place the Sun at the origin. To represent the orbiting object, draw a dot somewhere in the first quadrant, for example, and identify its coordinates (x, y). Connect the Sun at the origin to the dot with a vector **r** from the Sun to the dot and label the angle θ between the $+x$-axis and the line to the dot. Also show the attractive force acting on the dot by drawing a force vector **F** pointing from the dot toward the Sun. Resolve this force into components parallel to the x- and y-axes, then express these components in terms of **F** and θ. Next, express $\cos\theta$ and $\sin\theta$ in terms of the magnitude of **r** and the coordinates of the dot, (x, y). Finally, note that the mass of the comet, m, cancels when we equate the force to $m\mathbf{a}$.

With the help of your drawing, at this point you should easily understand the following assignments and equations:

In[27]:= **Fx = -G $\frac{Ms}{r^3}$ x[t]**

Fy = -G $\frac{Ms}{r^3}$ y[t]

Our **r** in Mathematica symbolizes the magnitude of our vector **r**. Hopefully you recall how to find the magnitude of a vector.

In[29]:= **r = $\sqrt{\{x[t], y[t]\} . \{x[t], y[t]\}}$**

We apply Newton's second law:

In[30]:= **eqns1 = {vx'[t] == Fx, vy'[t] == Fy}**

Since we want the position and well as the velocity of the comet, we include in our system of differential equations the two first-order differential equations relating the positions and the velocities. Thus

In[31]:= **eqns2 = {x'[t] == vx[t], y'[t] == vy[t]}**

With regard to units, we will choose to use one solar mass as the mass unit, astronomical units (AU.) as the distance unit, and years as the time unit. In this system

In[32]:= **G = 39.6; Ms = 1;**

Next we need some initial conditions. We will start the comet at aphelion where we know from astronomical data that

In[33]:= **initpos = {x[0] == 0.5871, y[0] == 0}**
  initvel = {vx[0] == 0, vy[0] == 11.52}
  {x[0], y[0]} == {.5871, 0}
  initveloc = {vx[0], vy[0]} == {0, 11.52}

where the distance is in AU and the velocity is in AU per year. We combine our equations

In[37]:= **halley = Union[eqns1, eqns2, initpos, initvel]**

and solve them:

In[38]:= **sol = Flatten[NDSolve[halley, {x, y, vx, vy}, {t, 0, 77}]]**

In[39]:= **ParametricPlot[{x[t], y[t]} /. sol, {t, 0, 77}, Compiled –> False, AspectRatio –> Automatic, AxesLabel –> {"x", "y"}];**

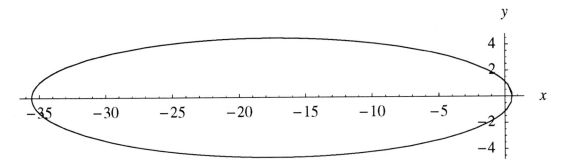

Kowabunga!

We prepare for some problems for you to solve by defining some quantaties we will need.

A. The velocity and its magnitude:

In[40]:= **v[t_] = {vx[t], vy[t]} /. sol**

In[41]:= **magv[t_] = $\sqrt{\text{v[t].v[t]}}$**

B. The vector **r** and its magnitude:

In[42]:= **rvector[t_] = {x[t], y[t]} /. sol**

In[43]:= **r /. sol**

C. The kinetic energy per unit mass:

In[44]:= **K[t_] = 1/2 v[t].v[t]**

D. The potential energy per unit mass:

In[45]:= **U[t_] = $-\dfrac{\text{G Ms}}{\text{r}}$ /. sol**

E. The angular momentum per unit mass and its magnitude:

In[46]:= **L[t_] = Cross[{x[t], y[t], 0}, {vx[t], vy[t], 0}] /. sol**

In[47]:= **magL[t_] = $\sqrt{\text{L[t].L[t]}}$**

Problem 10. Kepler's third law states that the square of the period T is equal to the cube of the semimajor axis a of the ellipse. Find T and a from the tools we have provided, and check this law.

Problem 11. Make and interpret the following graphs: **K[t]** versus **t**, **U[t]** versus **t**, **K[t] + U[t]** versus **t**, **magL[t]** versus **t**. Are both the total energy and the angular momentum conserved? Be sure to take note of the scale on the vertical axis.

8.3.2 Binary Stars

In[48]:= **Clear["Global`*"]**

How must we modify our approach if we are dealing with two objects of more or less equal masses? In this section we will provide the basic equations, leaving most of the Mathematica work up to you. In the next figure find the details we must draw to get all of the housekeeping straight. We have stars A and B at coordinates (x_A, y_A) and (x_A, y_A),

8.3 Motion Under the Influence of Gravity

respectively, with masses m_A and m_B. The gravitational force of star A on B is $\mathbf{F}_{AB}$ while the force of B on A is $\mathbf{F}_{BA}$. These forces are equal in magnitude to

$$G \frac{M_A M_B}{r^2} \qquad (4)$$

but, of course, opposite in direction. We will use the same units as before in which $G = 39.6$, distances are in AU., and time is in years.

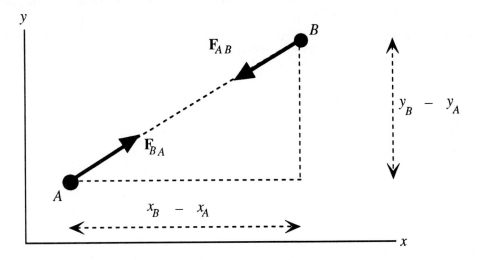

Here is Newton's second law for star A's x-component of motion.

$$M_A \frac{dv_{Ax}}{dt} = -G M_A \frac{M_B(x_A - x_B)}{r^3} \qquad (5)$$

where

$$r = \sqrt{(x_B - x_A)^2 + (y_B - y_A)^2} \qquad (6)$$

Similarly, the equation for the y-component of motion is

$$M_A \frac{dv_{Ay}}{dt} = -G M_A \frac{M_B(y_A - y_B)}{r^3} \qquad (7)$$

Problem 12. Write the equation of motion in both coordinates for star B. Use the fact that the force on star B must act in the opposite direction to that on star A.

Since we want positions as well we need to include the following equations:

$$\frac{dx_A}{dt} = v_{Ax}, \quad \frac{dy_A}{dt} = v_{Ay}, \quad \frac{dx_B}{dt} = v_{Bx}, \quad \frac{dy_B}{dt} = v_{By} \qquad (8)$$

We also need initial conditions; here are some suggested values:

$$x_A(0) = 10, \ y_A(0) = 0, \ x_B(0) = -10, \ y_B(0) = 0,$$
$$v_{Ax}(0) = 0, \ v_{Ay}(0) = 0.7, \ v_{Bx}(0) = 0, \ v_{By}(0) = -0.7 \quad (9)$$

Finally, you might try integrating over a time interval of 40 years. To begin, assume both masses are one solar mass, but try some other values once you get started.

Problem 13. OK, it's up to you. Translate these equations into the syntax of Mathematica and enter it into **NDSolve**. Graph the orbits of both stars using **ParametricPlot**. Use the models in previous sections to help you. When you get your model working, also give each star a little additional initial velocity in the y-direction and get them to move through space together. Experiment with stars of unequal masses. Don't expect everything to run the first time. Good luck!

8.3.3 Gravity-Assist Trajectories

In[49]:= **Clear["Global`*"]**

When sending spacecraft to Jupiter and other planets, NASA scientists use nearer planets to slingshot the spacecraft to higher speeds, greatly reducing the time it takes to get to the next planet on the journey. *Voyager I* and *II* were two outstanding example of this and, as we write, the spacecraft Cassini is making its way past several planets via a gravity-assist trajectory. In this section we illustrate a simple case of this process. We simulate the motion of a spacecraft past Mars, using it to pick up kinetic energy. What follows is complex in terms of housekeeping, but it is not any more profound than earlier topics. We assume all of our objects are in the same plane, which is only an approximation.

Although we need only consider the motion of one object, our spacecraft, it has the force of gravity of three different objects acting on it. And, although by virtue of its large mass the Sun may be considered fixed at the origin of our coordinate system, Earth and Mars move. However, rather than solve the differential equations of these two planets, we will assign them motions very much like their real motion. Let's begin at that point.

It is convenient to use parametric or vector equations to describe the position of Earth and Mars. Recall that in Mathematica $a\mathbf{i} + b\mathbf{j}$ = {a, b}. In particular, we will use the Mathematica expression

In[50]:= **re[t_] = {Sin[2π t], −Cos[2π t]}**

to describe the position of the Earth as a function of time. Observe that this is a circle of radius 1 AU and which completes an orbit when $t = 1$ year. Also observe that at $t = 0$, **re** = {0, −1}. The orbit of Mars is only slightly more difficult:

In[51]:= **rm[t_, θ_] = am $\left\{\text{Cos}\left[2\pi \dfrac{t}{tm} - \theta\right], \text{Sin}\left[2\pi \dfrac{t}{tm} - \theta\right]\right\}$**

8.3 Motion Under the Influence of Gravity

The **am** parameter is the mean radius of Mars' orbit, which is 1.52 a.u., and the **tm** parameter is the period of revolution of Mars, 1.882 years. We make the quantity θ a variable because we wish the spacecraft and Mars to arrive at almost the same point at the same time, and θ allows us to vary the time at which Mars arrives at any point. Notice that at $t = 0$

In[52]:= **rm[0, θ]**

Thus, θ determines the precise starting position of Mars. Mars must be very close to the spacecraft when it flies by if the spacecraft is going to acquire energy. We let θ control the position of Mars in its orbit. Let's plot parts of both of these orbits to illustrate.

In[53]:= **ParametricPlot[{re[t], rm[t, π/6] /. {am –> 1.52, tm –> 1.882}}, {t, 0, 1},
         PlotRange –> All, AspectRatio –> Automatic, Compiled –> False,
         AxesLabel –> {"x", "y"}];**

Finally, let the spacecraft be located with position vector **r** where

In[54]:= **r[t_] = {x[t], y[t]}**

Now would be a good time to use the graph we just plotted to draw a picture of the situation. Locate the Sun at the origin, Earth somewhere on its orbit with coordinates (x_E, y_E), Mars somewhere on its orbit with coordinates (x_M, y_E), and the spacecraft at arbitrary coordinates (x, y). Draw force vectors from the spacecraft toward the Sun, Earth, and Mars. Now we must express these forces in terms of the coordinates, but before we do that let's remember that when we use Newton's second law, the mass of the spacecraft cancels. So we will actually be giving forces per unit (spacecraft) mass. Refer to the figure you have drawn to vereify the following relationships.

A. The *x*-component of the force of the Sun on the spacecraft:

In[55]:= $\mathbf{Fsx\,[t\_]} = -\dfrac{G\,Ms\,x[t]}{(r[t].r[t])^{\frac{3}{2}}}$

B. The *y*-component of the force of the Sun on the spacecraft:

In[56]:= $\mathbf{Fsy\,[t\_]} = -\dfrac{G\,Ms\,y[t]}{(r[t].r[t])^{\frac{3}{2}}}$

C. The *x*-component of the force of the Earth on the spacecraft:

In[57]:= $\mathbf{Fex[t\_]} = -\dfrac{G\,Me\,\{1,\,0\}.(r[t]-re[t])}{((r[t]-re[t]).(r[t]-re[t]))^{\frac{3}{2}}}$

(Recall that the *x*-component of a vector may be found by dotting it with the unit vector **i**. In our case this means forming the dot product with {1, 0}.)

D. The *y*-component of the force of the Earth on the spacecraft:

$$\text{In[58]:= Fey[t\_] = } - \frac{G \, Me \, \{0, 1\} . (r[t] - re[t])}{((r[t] - re[t]) . (r[t] - re[t]))^{\frac{3}{2}}}$$

E. The *x*-component of the force of Mars on the spacecraft:

$$\text{In[59]:= Fmx[t\_, } \theta\text{\_] = } - \frac{G \, Mm \, \{1, 0\} . (r[t] - rm[t, \theta])}{((r[t] - rm[t, \theta]) . (r[t] - rm[t, \theta]))^{\frac{3}{2}}}$$

F. The *y*-component of the force of Mars on the spacecraft:

$$\text{In[60]:= Fmy[t\_, } \theta\text{\_] = } - \frac{G \, Mm \, \{0, 1\} . (r[t] - rm[t, \theta])}{((r[t] - rm[t, \theta]) . (r[t] - rm[t, \theta]))^{\frac{3}{2}}}$$

We assemble the differential equations in preparation for **NDSolve**.

```
In[61]:= diffeqs = { vx'[t] == Fsx[t] + Fex[t] + Fmx[t, θ],
                    vy'[t] == Fsy[t] + Fey[t] + Fmy[t, θ]}
         morediffeqs = {x'[t] == vx[t], y'[t] == vy[t]}
```

We also assemble the initial conditions we will use. Note that we cannot have the spacecraft start exactly at the center of the Earth because that makes the gravitational force of the Earth infinite. So we start the spacecraft at {0.001, −1}, whereas the Earth is at {0, −1}.

```
In[63]:= initcond = {vx[0] == 7.078135, vy[0] == 0, x[0] == 0.001, y[0] == −1}
```

Finally, we collect the parameters and variable values we will need:

```
In[64]:= G = 39.6; Me = 3.01 10^−6; Mm = 3.23 10^−7; Ms = 1; θ = .783181;
         am = 1.52; tm = 1.882;
```

The solution to this system of differential equations follows, but first let's explain how we found the values for the initial condition **vx[0]** and the variable θ. The best approach is to put the spacecraft in a so-called *transfer* orbit from Earth to Mars. This means that the spacecraft is launched when Earth is at perhelion and arrives at Mars when it is at aphelion. This minimizes the energy required to send a spacecraft to Mars. Since our Earth and Mars are in hypothetical circular orbits in our model, aphelion and perhelion are at the same distance. We launched our spacecraft so that it would start at the Earth and end at Mars, and we found this by trial and error; that is, by solving the differential equation over and

8.3 Motion Under the Influence of Gravity

over until the spacecraft intersected the orbit of Mars just when Mars would be turning away from it.

Once this was achieved with **vx[0] = 7.078135**, we adjust θ by trial and error to have Mars and the spacecraft arrive at the almost the same place at the same time. In particular, we made a graph of the kinetic energy of the spacecraft as a function of time, and adjusted the angle θ to get a jump in the kinetic energy during the flyby. These can be tricky, extremely time-consuming adjustments, and we do not guarantee that we have achieved the maximum energy jump possible, but our approach is similar to that of the experts. They program many different orbits and then choose the one that gives the desired result.

Here is the solution:

In[65]:= **sol = NDSolve[Union[diffeqs, morediffeqs, initcond], {x[t], y[t], vx[t], vy[t]}, {t, 0, 2}, WorkingPrecision -> 18]**

This instruction makes a graph of the orbits of Earth, Mars, and the spacecraft, but does not display it.

In[66]:= **g1 = ParametricPlot[**
  {rm[t, θ], re[t], {x[t], y[t]} /. sol}, {t, 0, 2}, PlotRange -> All,
  PlotPoints -> 400, AspectRatio -> Automatic, Compiled -> False,
  DisplayFunction -> Identity, AxesLabel -> {"x", "y"}];

These commands calculate the kinetic energy and its graph without a display.

In[67]:= **e[t_] := 1/2 (vx[t]^2 + vy[t]^2)**
  g2 = Plot[e[t] /. sol, {t, 0, 2}, PlotRange -> All, DisplayFunction -> Identity,
  AxesLabel -> {"t", "K"}];

This command displays both the orbits and the graph of the kinetic energy.

In[69]:= **Show[GraphicsArray[{g1, g2}]];**

The graph at the head of the chapter is similar to what you should get. While searching for a good solution, we also found it essential to plot tiny sections of the orbits of both Mars and the spacecraft near their point of near collision. This plot helped.

In[70]:= **g3 = ParametricPlot[{rm[t, θ], {x[t], y[t]} /. sol}, {t, 0.70, .71},**
  PlotRange -> All, PlotPoints -> 400, Compiled -> False];

Problem 14. With one or more classmates, discuss what the conservation of energy means in the situation just described. The kinetic energy of the spacecraft is certainly not conserved. What energy is conserved?

Problem 15. Also discuss what you think is the best point in the orbit of Mars to have the flyby occur. Does it matter? Does the reality of non-circular orbits change whether or not it matters?

Problem 16. Finally, discuss the fact that as the spacecraft approaches Mars, it picks up kinetic energy as it falls toward Mars. Should it not lose the same amount of kinetic energy (as immediately as it gained it) as it leaves Mars, leaving its total kinetic energy unchanged?

Problem 17. Do the gravitational effects of the Earth make a profound influence on the orbit of the spacecraft?

Problem 18. This takes some trial and error and with it time, but increase the mass of Mars by an order of magnitude and repeat the experiment. Can you get a greater change in the kinetic energy? You must experiment with the initial velocity and the phase angle to optimize the kinetic energy change.

■ 8.4 An Example from Biology

In[71]:= **Clear["Global`*"]**

Many biology students take courses in physics; perhaps you are one of them. Differential equations also arise in biology. One of my colleague's favorite sayings to his differential equations class is; Life is a differential equation. However, no one knows what the differential equation is, and we will now look at a much simpler system of differential equations from biology, namely the Lotka-Volterra predator-prey equations. This problem occurs under many names, foxes and rabbits, wolves and caribou, but Lotka's concern was originally with a herbivorous animal population that preyed on some plant populations; we will choose wolves and caribou for our study. Don't take our results very seriously, they are only intended to show some concepts, and they do not represent any sort of actual study. Clearly the wolves are the predators and the caribou are the prey. Without predators and an unlimited food supply on the tundra, the rate of change of caribou would be

$$\frac{dc}{dt} = Ac \tag{10}$$

Thus the rate of change of caribou is proportional to the existing population of caribou. However, with some wolves around, the rate of change of caribou needs another term, namely

$$\frac{dc}{dt} = Ac - Bcw \tag{11}$$

The product cw represents the number of caribou-wolf encounters, and the constant B depends on what fraction of them result in a fatality for the caribou.

Now we need a differential equation for the wolves, and here it is:

8.4 An Example from Biology

$$\frac{dw}{dt} = Gcw - Hw \qquad (12)$$

The *Gcw* term shows that what is bad for the caribou-wolf encounters is good for the wolves, while the *Hw* term, being negative, means that wolves are not immortal themselves. In these equations, *A*, *B*, *G*, and *H* are assumed to be constants. Of course, a more complex model might have these values changing with other factors in the environment – human beings, for example, the deadliest predator of all. We set up these equations in Mathematica:

In[72]:= **diffeq = {caribou'[t] == A caribou[t] – B caribou[t] wolves[t],
         wolves'[t] == G caribou[t] wolves[t] – H wolves[t]}**

As always, we need some initial conditions, and without any knowledge of a real situation we try the following:

In[73]:= **initcond = {wolves[0] == 200, caribou[0] == 800}**

Is there a stable situation where the number of wolves and the number of caribou remain constant? In that case, the derivatives are zero and the quantities keep their initial values. We try the following command

In[74]:= **Solve[{0 == A 800 – B 800 200, 0 == G 800 200 – H 200}, {A, B, G, H}]**

fully realizing that we have too many unknowns.

Eureka! We find that if A/B = wolves[0] = 200 and G/H = caribou[0] = 800 we have stability. Thus if $A = 2$, $B = 0.01$, and if $G = 0.01$, then $H = 8$. Relying mostly on ignorance, we go ahead and assign these values:

In[75]:= **A = 2; B = 0.01; G = 0.01; H = 8;**

We should know what the answer is going to be, but we go ahead and solve the differential equations.

In[76]:= **sol = NDSolve[Union[diffeq, initcond], {caribou[t], wolves[t]}, {t, 0, 10}]**

Plot it if you wish, but you know what to expect.

In[77]:= **Plot[{wolves[t] /. sol, caribou[t] /. sol}, {t, 0, 10}];**

We wish to experiment with the wolf-caribou ecology when, say, 100 additional wolves are imported into this system. We must modify the initial conditions which we choose to do as follows:

In[78]:= **sol = NDSolve[Union[diffeq, {wolves[0] == 300, caribou[0] == 800}],
         {caribou[t], wolves[t]}, {t, 0, 10}]**

In[79]:= **Plot[{wolves[t] /. sol, caribou[t] /. sol}, {t, 0, 10},**
          AxesLabel -> {"t", "Wolves/Caribou"}];

Aha! The populations oscillate, which probably does not surprise us either. Here is another way of viewing our results.

In[80]:= **ParametricPlot[{wolves[t], caribou[t]} /. sol, {t, 0, 10}, Compiled -> False,**
          AxesLabel -> {"Wolves", "Caribou"}];

Problem 19. What happens to the number of caribou if there are no wolves? What happens to the wolves if there are no caribou? Make some predictions of your own before you solve the differential equations.

Problem 20. Suppose the caribou herd is reduced to 80 members because caribou suddenly became popular in resturants, while the number of wolves remains at 200. What happens to the number of wolves? Which population, if any, do you predict will end at zero? What happens to the number of caribou? Does the solution to the differential equations reflect what really happens at all subsequent times? Remember, the differential equation does not realize it takes at least two wolves of opposite sex to reproduce.

Mostly Mathematica

1. Use **Table** to make a list of coordinates of $\{n, n^2\}$ for n from 1 to 30. **ListPlot** this table. What kind of a curve do you get? If you have what Mathematica calls a list, which is just a set, some functions distribute themselves over the list. **Log** is such a function. Distribute **Log** over the list provided by your **Table** command, then make a **ListPlot** of this table. How does this curve differ from the previous one?

2. Use Mathematica's **Sum** (or $\sum_{\square=\square}^{\square} \square$ from the palette) to add the first 1000 counting numbers. You may need to perform a **??Sum** command to find the correct syntax.

3. The average value of a continuous function over an interval $[0, T]$ is

$$\frac{1}{T} \int_0^T f(t) \, dt \tag{13}$$

Find the average value of $\sin t$ and $\sin^2 t$ over one cycle, $[0, 2\pi]$.

Explorations

1. Try to model a gravity-assist trajectory past Jupiter.

2. In Section 8.2, we modeled an elastic collision. Think of a way to model an inelastic collision and do this in Mathematica.

3. In Chapter 4, we used our own numerical integration technique to make a crude model of an *Apollo* mission. Solve the same problem using **NDSolve**.

4. Introduce a third star into the binary star problem, and have the two orbiting stars travel through space to collide with the third star. Our results for this problem are shown as the first figure in Chapter 7.

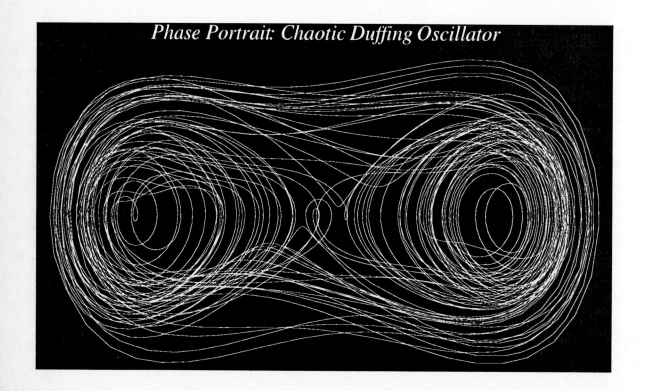

Phase Portrait: Chaotic Duffing Oscillator

CHAPTER 9 Oscillatory Motion

■ 9.1 Introduction

The world abounds with examples of oscillations: butterfly wings, leaves, atoms in molecules, cars bouncing on roads, radio transmitters, tornados, animal populations, wings on airplanes. As I write this chapter, the 17-year locusts have come out of the ground to fly around and reproduce; this represents an oscillation with a period of 17 years. Each locust can sing, which makes a second locust oscillation. Finally, later in the afternoon, the locusts in a single tree or group of trees appear to synchronize their singing, and together they modulate their song so the whole environment around the house ebbs and swells with what sounds like a King-Kong locust. In fact, except for my curiosity about this happening, I will be glad when they are gone. In any case, some oscillations are extremely difficult to analyze mathematically; others are quite simple. We will look at a few simple cases.

9.2 The Simple Harmonic Oscillator

In[1]:= **Clear["@"]**

We consider a mass *m* connected to a spring with spring constant *k*. The mass is resting on a horizontal frictionless surface. Our coordinate system is horizontal with the origin at the position of the center of mass when the spring is unstretched. The force of the spring on the mass is $-kx(t)$, where $x(t)$ is the displacement of the mass from the unstretched position as a function of time. We wish to find the position, $x(t)$, velocity, $v(t) = x'(t)$, and acceleration, $a(t) = x''(t)$ of the mass.

Newton's second law applied to the spring-mass system gives

$$-kx(t) = mx''(t). \qquad (1)$$

Not wishing to say that this equation is a differential equation and needs to be solved using rather sophisticated mathematics, at this point your text may say instead: Try a solution of the form ... , pulling a proposed solution, which you are to substitute into Equation (1), out of more-or-less thin air. We will not improve much on that situation. We will appeal to Mathematica's **DSolve** command to find the solution. In its simplest form, this command is

In[2]:= **DSolve[m x''[t] == −k x[t] , x[t], t]**

Using the command in this form gives us two constants of integration, *C*[1] and *C*[2], which we must determine from the initial conditions. It is more efficient to enter the initial conditions directly into the **DSolve** command. Let's assume that the mass is simply pulled to a position $x = 1$ and released. This gives the initial conditions: **x[0] ==1** and **v[0] == 0**. Here is the solution:

In[3]:= **sol = DSolve[{m x''[t] == −k x[t], x[0] == 1, x'[0] == 0} , x[t], t]**

We should not be too surprised to find the answer is a sine or cosine function. Finally, since we will typically want **v[t]** as well, we modify the arguments of the **DSolve** command as follows:

In[4]:= **solution = DSolve[{m v'[t] == −k x[t], x'[t] == v[t], x[0] == 1, v[0] == 0} , {x[t], v[t]}, t]**

Problem 1. Carefully describe how does the solution change if the initial conditions are changed to **x[0] == 0** and **v[0] ==1**? What happens if the initial conditions are **x[0] ==1** and **v[0] == 1**?

From the solution and your knowledge of trig functions, you can determine the period. Since cosine is periodic in 2π, one period, *T*, of the cosine function will have elapsed when

$$\sqrt{\frac{k}{m}}\, T = 2\pi \qquad (2)$$

or $T = 2\pi\sqrt{\frac{m}{k}}$. In that case we also know that the frequency is given by

$$f = \frac{1}{T} = \frac{1}{2\pi}\sqrt{\frac{k}{m}} \qquad (3)$$

To plot the solution, we must assign values to k and m. To begin, we assume

In[5]:= **k = m = 1;**

In[6]:= **Plot[{x[t] /. solution, v[t] /. solution}, {t, 0, 4 π},
AxesLabel –> {"t", "x/v"}];**

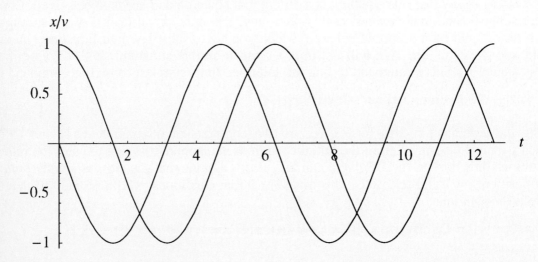

We can also find the acceleration:

In[7]:= **a[t_] = ∂_t (v[t] /. solution)**

We can plot all three functions on the same coordinate system as follows:

In[8]:= **Plot[{x[t] /. solution, v[t] /. solution, a[t]}, {t, 0, 2 π},
AxesLabel –> {"t", ""}];**

Problem 2. What effect on the period does increasing the mass have? Does this make sense when you think of mass as measuring inertia? Explain.

Problem 3. Predict what effect increasing the spring constant has on the period. Then replot the solution for several values of k and verify your prediction.

9.3 The Duffing Oscillator

Problem 4. Instead of graphing the three curves on the same coordinate axes, make a graphics array of the three graphs. Use the commands **Show** and **GraphicsArray**.

Problem 5. Another useful way to plot the results is to use *phase-space* with x along the horizontal axis and v along the vertical axis. We must use **ParametricPlot** to do this. To introduce some variety, we change k and m.

In[9]:= **k = 1; m = 1/2**

In[10]:= **ParametricPlot[{x[t], v[t]} /. solution, {t, 0, 2 π}, AxesLabel -> {"x", "v"},
         Compiled -> False];**

Interpret this graph. Where is the mass on the spring when the curve crosses the +v-axis? Where is the mass when the curve crosses the −x-axis? Why does the graph have the shape of an ellipse? Is it an ellipse or does it simply look like an ellipse?

Problem 6. Recall that

$$U(x) = -\int F(x)\,dx. \qquad (4)$$

Use Mathematica to find and graph the potential energy function for the mass on a spring. For your graph assume $k = 1$, and let $U(0) = 0$.

Problem 7. Is energy conserved during the oscillation? Find the sum of the kinetic and potential energies of the mass and plot this sum as a function of time.

■ 9.3 The Duffing Oscillator

In[11]:= **Clear["Global`*"]**

Another system we will study, which is probably not found in your textbook, is the so-called Duffing oscillator. The oscillator we will study in particular has a potential energy (per unit mass) function

In[12]:= **U[x_] = $-\dfrac{x^2}{2} + \dfrac{x^4}{4}$**

and a force (per unit mass) function

In[13]:= **F[x_] = $-\partial_x$ U[x]**

Problem 8. Make a graph of **U[x]** and **F[x]** on the same coordinate system.

Problem 9. Interpret these graphs in terms of the behavior of a particle placed in such a potential well with some initial position and velocity.

Next we try to solve the Duffing oscillator equation of motion for a particle whose $m = 1$ with **DSolve**. Here is what happens:

In[14]:= **DSolve[{v'[t] == x[t] – x[t]^3, x'[t] == v[t], x[0] == 1, v[0] == 1},
      {x[t], v[t]}, t]**

DSolve fails, so we try a numerical solution with **NDSolve**.

In[15]:= **solution = NDSolve[{v'[t] == x[t] – x[t]^3, x'[t] == v[t], x[0] == 1, v[0] == 1},
      {x, v}, {t, 0, 10}]**

We graph the results in a phase-space plot.

In[16]:= **ParametricPlot[{x[t], v[t]} /. solution, {t, 0, 10},
      AxesLabel –> {"x[t]", " v[t]"}, Compiled –> False];**

Problem 10. Interpret the phase-space graph of the Duffing oscillator. In your interpretation be sure to refer to the potential energy diagram you made earlier and mention the effects of the small hill in the potential energy. Try some different initial conditions such as **x[0]==1, v[0] ==–0.5** or **x[0] ==0.1, v[0]==0** and explain your results.

■ 9.4 Damped Oscillations

In[17]:= **Clear["Global`*"]**

Your car represents a mass on a spring, but it does not continue to oscillate forever after it hits a bump. Aside from the fact that there are energy losses in the springs, most cars have shock absorbers to damp out the oscillations. If we were to mount a vane on our oscillating mass on a spring, then the mass would experience an additional force proportional to the negative of the velocity, which we identify as $-cv(t)$. The constant c depends on the vane size and the density of air, among other factors. We insert this new term into Newton's second law.

In[18]:= **solution =
      DSolve[{m v'[t] == –k x[t] – c v[t], x'[t] == v[t], x[0] == 1, v[0] == 0},
      {x[t], v[t]}, t]**

9.6 Tracking the Duffing Oscillator on the Road to Chaos

In[27]:= **Table[Plot[y[t] /. ω -> w, {t, 950, 1000}, GridLines -> Automatic],
{w, .5, 1.5, .1}]**

Problem 14. After you have finished all the graphs, make a table by hand that shows the amplitude of the vibration as a function of the driving frequency ω. Then make a graph of amplitude versus ω. Resonance occurs when $\omega = \omega_o$. What happens to the amplitude when that is true.

Problem 15. Repeat the process just described for a larger value of γ, say $\gamma = 0.1$. What happens to the resonance curve, amplitude versus ω, as γ gets larger?

■ 9.6 Tracking the Duffing Oscillator on the Road to Chaos

We wish to continue our study of damped and driven oscillations by returning to the Duffing oscillator. Recall that we have thought of the Duffing oscillator in terms of a particle rolling around in the double well of the potential function for this oscillator. That is acceptable as long as we have nothing more complex than our ficticious particle in the well; however, adding damping causes a minor complication; how do we damp such a particle? We could add a vane, of course. A more difficult a question is: how do we drive such a particle? Perhaps we could put our ficticious frictionless mass (with vane) in such a well, and then place the entire apparatus in a gravitational field on a table that is oscillating up and down at the frequency of the driving force. Or perhaps it is simpler to just think of gravity varying. Or perhaps we need not worry about it and just assume it can be done. This will be our approach.

We make a fresh start with

In[28]:= **Clear["@"]**

and then we define our parameters for damped and driven oscillations.

In[29]:= **γ = 0.1; d = 0.1; ω = 1.4;**

Notice that in the **NDSolve** command we will add a driving term, **d Cos[ωt]**. In what follows, we will want our solutions to cover a considerable amount of time, so we increase the default number of steps allowed by Mathematica with a **MaxSteps** option. We also start the particle at $x = v = 0$.

In[30]:= **solution = NDSolve[{v'[t] == x[t] - x[t]^3 - γ v[t] + d Cos[ω t],
x'[t] == v[t], x[0] == 0, v[0] == 0}, {x[t], v[t]}, {t, 0, 200},
MaxSteps -> 2000]**

We graph the results in a phase-space plot. Let's begin by graphing the results over half of our original interval; it may be easier to see the evolution of the oscillations that way.

In[31]:= **ParametricPlot[{x[t], v[t]} /. solution, {t, 0, 100},**
 AxesLabel -> {"x(t)]", " v(t)"}, Compiled -> False, PlotPoints -> 100];

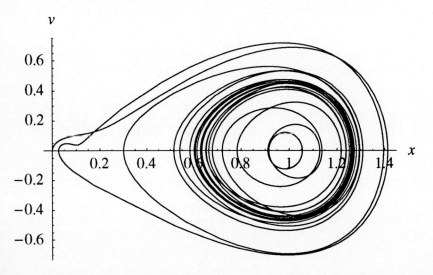

Now we graph the trajectory in phase-space over the last fourth of the interval our solution covers.

In[32]:= **ParametricPlot[{x[t], v[t]} /. solution, {t, 150, 200},**
 AxesLabel -> {"x(t)", " v(t)"}, Compiled -> False, PlotPoints -> 100];

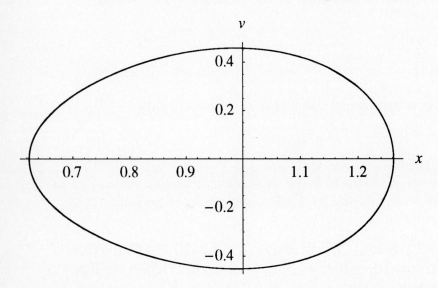

9.6 Tracking the Duffing Oscillator on the Road to Chaos

Here is our interpretation of these graphs. It seems clear that the particle begins in the central potential local maximum of the Duffing oscillator at $x = 0$ with no velocity, corresponding to our initial conditions, but the driving force moves it to the right. Although it may wander around in phase-space to begin, it appears to settle into a stable orbit (called a *limit cycle*) during the last part of the interval. Because the particle has insufficient energy, the orbit never crosses over the hill in the center to get to the other well.

Problem 15. Find the period of the oscillations. Use trial and error and **ParametricPlot** to plot the solution over one cycle. Is the period you found, 4.49 s, the same as $2\pi/\omega$ where $\omega = 1.4$, the period of the driving force?

We are going to solve the differential equation and graph the solution several times with different values for the driving force amplitude, d. It is wise, in this case, to make functions to accept d as a variable, as follows:

In[33]:= **Clear["@"]**

We will keep γ and ω the same.

In[34]:= **γ = 0.1; ω = 1.4;**

In[35]:= **solution[d_, tmax_] := Flatten[
    NDSolve[{v'[t] == x[t] − x[t]^3 − γ v[t] + d Cos[ω t],
    x'[t] == v[t], x[0] == 0, v[0] == 0},
    {x[t], v[t]}, {t, 0, tmax}, MaxSteps −> 10000]];**

> We have increased **MaxSteps** to 10,000 to be able to find a solution up to **tmax** = 500. This requires quite a bit of memory. Be aware, and frequently **Save** your results. You may have to restart the kernal if you run out of memory.

After assigning $d = 0.32$, we get a solution this way:

In[36]:= **sol = solution[0.32, 500]**

Here is a function we created that will graph the solution over any time interval up to **tmax**.

In[37]:= **graph[tmin_, tmax_] := ParametricPlot[{x[t], v[t]} /. sol, {t, tmin, tmax},
    AxesLabel −> {"x[t]", " v[t]"}, Compiled −> False, PlotPoints −> 100];**

We can graph whatever portion of the solution we want as follows:

In[38]:= **graph[0, 250];**

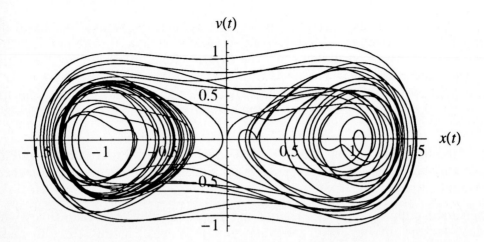

This looks like a terrible mess, but perhaps the oscillator is just settling down. Let's try it over a much smaller interval, preferably toward the end in the hope that things have settled down. Before doing that, we should note that the Duffing oscillator particle has moved through both of the wells.

  In[39]:= **graph[450, 500];**

Once again, the oscillator appears to be in a limit cycle; after some transitory behavior where it wanders around in phase space, it settles down and finds a stable orbit. However, if we look at this orbit carefully, we will notice that to complete an orbit it travels *twice* around the potential minimum at $x = -1$.

Problem 16. Use trial and error and the parametric plotting function we created to find the new period of oscillation. Is it twice $2\pi/\omega$?

The phenomenon we have just observed is called *period doubling*. Initially the system has some tendency to oscillate at the natural frequency of the system, but these natural oscillations die out, and the system oscillates at the frequency of the driving force. Increasing this force causes the oscillations to change to have a period twice the period of the driving force, then four times the period of the driving force, and so on until another dramatic change takes place and the system becomes *chaotic*. In this latter state there is no periodicity whatsoever; the orbit never closes on itself. To illustrate, we increase the driving force to 0.35 and find and graph a new solution.

  In[40]:= **Clear[sol]**
           sol = solution[0.35, 900]

  In[42]:= **graph[200, 900];**

You can see part of the graph at the head of this chapter. We will continue our study of chaos in Chapter 13. We need a much more careful definition of chaos than a phase-space trajectory that looks chaotic, but we assure you that the motion we have described is indeed chaotic.

9.7 The Lorenz Equations

In[43]:= **Clear["@"]**

We have not yet illustrated the solution of a system of differential equations in three dimensions, and we will briefly do so. The three equations we will use became famous as a result of the work of E. N. Lorenz in studying the atmosphere., Lorenz found that under certain conditions the system of equations is chaotic. As you might guess, this has a rather profound influence on how we regard weather and its prediction. In any case, here are the three equations

$$x'(t) = \sigma(y-x)$$
$$y'(t) = -xz + rx - y \qquad (8)$$
$$z'(t) = xy - bz$$

where σ, r, and b are parameters and x, y, and z are variables for some quantities of fluid dynamics. We express the same equations in Mathematica as follows:

In[44]:= **eqns = {x'[t] == σ (y[t] − x[t]),**
       y'[t] == −x[t] z[t] + r x[t] − y[t],
       z'[t] == x[t] y[t] − b z[t]}

Here are a set of parameters widely used to illustrate possible solutions.

In[45]:= **σ = 10; b = 8/3; r = 26.5;**

We solve this system of three coupled differential equations with **NDSolve**. You can identify the initial conditions from the argument of **NDSolve**.

In[46]:= **lorenz = NDSolve[Union[eqns, {x[0] == 1, y[0] == 0, z[0] == 0}],**
       {x, y, z}, {t, 0, 40}, MaxSteps −> 4000]

Since there are three variables, we use a three-dimensional plot to illustrate one possible trajectory in phase-space.

In[47]:= **ParametricPlot3D[Evaluate[{x[t], y[t], z[t]} /. lorenz], {t, 10, 40},**
       PlotPoints −> 4000, PlotRange −> All, AxesLabel −> {"x", "y" "z"}];

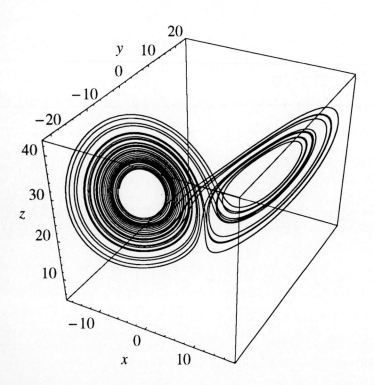

Problem 17. Make a graph of each of the three variables as a function of time. Do the solutions appear to be chaotic?

Mostly Mathematica

1. Find the area under the graph of

$$f(v) = v^2 e^{-v^2} \qquad (9)$$

on the interval $[0, \infty)$. Begin by graphing the function. What is

$$\lim_{v \to \infty} f(v) \qquad (10)$$

At what value of v is the function a maximum?

2. Use **FindRoot** to find all of the zeros of the function $f(x) = \frac{x}{2} - 2\cos(x)$. Begin by defining the same function in Mathematica and graphing it.

3. Find the real solutions to this system of equations: $y = x^2$ and $x^2 + y^2 = 1$. Graph both equations.

Explorations

1. Use **NDSolve** to study the motion of the simple pendulum. Begin with the equation of motion

$$\theta''(t) = -\frac{g}{L} \sin(\theta((t))) \qquad (11)$$

but add a damping term, $-\gamma\theta'(t)$ and a driving term $d\cos(\omega t)$ once you get your commands working. Does the motion become chaotic? Can you observe period doubling? See also De Jong (1991), Giordano (1997), and Gould and Tobochnik (1988).

2. If you have some experience with electrical circuits, find and solve the differential equation for an *LRC* circuit driven by some ac source.

3. Think of a car as a mass on a spring with some kind of damping. What happens when the car hits a bump or a series of bumps? In this case we may consider the driving force to be a step function of some kind.

$$F(t) = 0,\ 0 \le t \le t_1,\ F(t) = 1,\ t_1 < t \le t_2 \qquad (12)$$

Use **NDSolve** to solve this problem and graph the results.

References

De Jong, M. L. *Introduction to Computational Physics*, Reading, Mass.: Addison-Wesley, 1991, p. 205.

Giordano, N. J. *Computational Physics*, Upper Saddle River, N.J.: Prentice Hall, 1997, p. 42.

Gould, H., and J. Tobochnik, *Computer Simulation Methods*, Reading, Mass.: Addison-Wesley, 1988, p. 152.

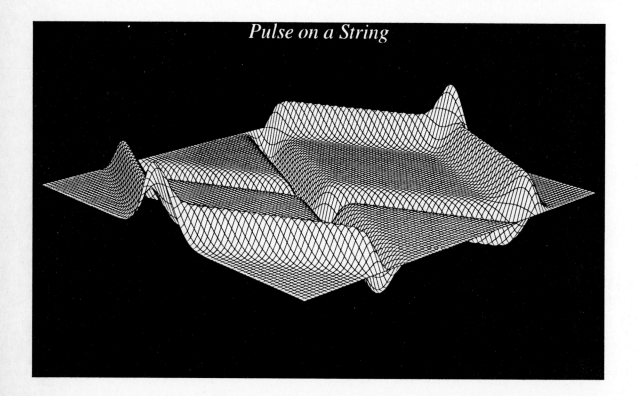

Pulse on a String

CHAPTER 10 Topics in Wave Motion

■ 10.1 Introduction

In this chapter we will look at a variety of topics typically covered either in chapters on wave motion, sound, or optics. Associated with every wave is some physical quantity that varies in time and space. In the case of sound waves, it is the pressure in the air that changes; in the case of light waves, it is the electromagnetic fields that change, while in the case of a water wave or a wave on a violin string, it is the displacement of a water molecule or rope particle from some equilibrium position that changes. Sometimes waves are more subtle, like the probability amplitudes of quantum mechanics. Common to these waves is the ability to express the quantity that varies as some sinusoidal function (or the sum of sinusoids) of time and position and the opportunity to use the superposition principle to combine waves that meet at some position in space. For example, the superposition principle guarantees that the **E**-fields associated with electromagnetic waves will add when two or more waves come together in space. Likewise, the pressure variations that accompany two sound waves will add. Thus much of the work we do below applies to a large variety of waves.

Before almost drowning in the abstraction that follows, it might be important to remind ourselves that our goal is to use the superposition principle to find the intensity of two or more waves with a phase difference ϕ coming together at some point and adding. Perhaps the most famous example of this is Young's double-slit experiment, but the same effect can be accomplished with sound waves and water waves. The sources of the waves might be an array of speakers, holes in a dike (perish the thought), slits in an opaque screen, layered reflecting surfaces, or a set of radio antennas. The phase difference ϕ can be produced electronically in the case of speakers and radio antennas, or it can be produced by making one wave travel a different distance than the other, which is what happens in Young's double-slit experiment. It is to be hoped that you have worked your way through one or more of these examples in your class before dealing with the mathematics of the problem using Mathematica.

The problem of finding the resultant amplitude usually begins with a trignometric sum such as

$$A_0 \sin(\omega t) + A_1 \sin(\omega t + \phi) + A_2 \sin(\omega t + 2\phi) + \ldots + A_n \sin(\omega t + n\phi) \tag{1}$$

Then, to find the intensity, the time average of the square of Equation (1) must be found. It is precisely in dealing with the mathematics of the problem that Mathematica is so useful. The typical textbook approach is to use various geometric and trigonometric techniques to form the sum and time average of a series of sine or cosine functions. Mathematica can handle the problem head-on without the introduction of phasors or tricky trigonometry.

∎ 10.2 The Mathematical Expression of a Wave

We consider a wave traveling along the +x-axis. We shall use the following expression to describe a simple wave:

$$y(x, t) = A \sin\left(2\pi \frac{x}{\lambda} - \omega t\right) \tag{2}$$

where x is a position coordinate, t is the time, λ is the wavelength, and ω is the angular frequency which is related to the frequency f by

$$\omega = 2\pi f \tag{3}$$

Note also that

$$\lambda f = v \tag{4}$$

where v is the wave velocity. The amplitude A can be a pressure amplitude, electromagnetic field amplitude, or a displacement depending on whether we are discussing a sound wave, electromagnetic wave, or wave on a string.

For the moment, assume we are discussing a traveling wave on a string so that A is the amplitude of the displacement of the string perpendicular to the undisturbed position of the string. Then Equation (1) describes how the displacement varies with x for some given time or how a particular particle of the horizontal string at a given position x moves up and down in time. We can plot the displacement of the string molecules as a function of position or as a function of time. To plot the wave as a function of time, we choose an arbitrary position such as the origin ($x = 0$) and make a graph of y as a function of time. Of course, Mathematica needs values for A, λ, and ω before it can proceed; for simplicity we choose unity for all three parameters.

In[1]:= **Clear["Global`*"]**
A = λ = ω = 1;
y[x_, t_] := A Sin$\left[2\pi \dfrac{x}{\lambda} - \omega t\right]$
Plot[y[0, t], {t, 0, 4π}, AxesLabel –> {"t", "y"}];

Or we can plot the displacement along the x-axis at a particular instant. Let's choose $t = 1$.

In[4]:= **Plot[y[x, 1], {x, 0, 3}, AxesLabel –> {"t", "y"}];**

Problem 1. How can you tell from this plot that the wavelength we chose is 1?

We can also plot the wave as a function both of x and t and obtain a picture of both its variation in space and in time. We use Mathematica's **Plot3D** command as follows:

In[5]:= **Plot3D[y[x, t], {x, 0, 3},**
 {t, 0, 3π}, PlotPoints –> 30, AxesLabel –> {"x", "t", "y"}];
 (*produces a color plot*)

In[6]:= **Plot3D[y[x, t], {x, 0, 3}, {t, 0, 2π},**
 Shading –> False, PlotPoints –> 40, AxesLabel –> {"x", "t", "y"}];
 (*produces a black & white plot*)

From the three-dimensional plot you can see both the variation of the displacement y with position x, and the variation of y with time for a molecule at a particular x. Don't think of this as some kind of two-dimensional ocean wave. Think about this figure carefully.

Finally, we can animate the wave as it travels to the right along the x-axis. We need to make a series of graphs of the displacement as a function of x for various times. The next command will do that. Then select all of the graphs and go to **Animate Selected Graphics** in the **Cell** menu of Mathematica. Adjust the speed to give the appearance of a wave traveling to the right.

In[7]:= **Table[Plot[y[x, t], {x, 0, 3}, AxesLabel −> {"t", "y"}], {t, 0, 4 π, .5}];**

> Animation requires quite a few graphs and therefore uses a lot of memory. After you are finished with the animation and the exercises in the next set, you might **Cut** the graphs from your notebook to preserve memory for Mathematica.

Problem 2. Either study the animated wave or the series of plots of y versus x at subsequent times, and suppose the wave is a traveling wave on a string. What is the *phase* of the string molecule at $x = 0$ at $t = 0$? That is, what is its displacement y and is it moving up or down? Specifying the displacement y and the rate of change of the displacement, dy/dt, specifies the phase. Two points on the wave having the same phase are one wavelength, λ, apart.

Problem 3. What is the displacement y and the sign of dy/dt when $x = 0$ and $t = \pi/4$?

Problem 4. From the animated wave or the series of plots, determine how far a crest goes in the time interval of the plots, namely 4π s. We found that a crest moved approximately 2 m. What does that give for the wave velocity?

Problem 5. Equation (3) also gives an expression for the wave velocity, $v = \lambda f$. Since we chose $\omega = 1$ and $f = \omega/2\pi f$, we can also find v from λf. Compare your results from Problems 4 and 5.

■ 10.3 Beats: A Simple Example of Interference

One of the more interesting phenomena associated with the combining of waves is what you hear when two sound waves of slightly different frequency arrive at your ears. Musicians use these beats to tune their instruments to the same frequency, and beats become annoying when traveling in a small twin-engine airplane and the engines are running at slightly different angular speeds. We take the origin of our coordinate system to be at the point where the waves from two hypothetical violins combine, so in our mathematical expression of the waves we take $x = 0$. We will also express the time variation of the waves in terms of frequency f rather than angular frequency ω. We give one wave a frequency of 256 Hz and the other a frequency of 300 Hz. They both have the same unit amplitude.

In[8]:= **Clear["@"]**
w1 = Sin[2 π 256 t];
w2 = Sin[2 π 300 t];

Here is a graph of the sum of the two waves. From the graph you can see the beats. The beat frequency is simply the difference in the frequencies 300 Hz – 256 Hz.

In[10]:= **Plot[w1 + w2, {t, 0, .1}, AxesLabel –> {"t", "w"}, PlotPoints –> 50];**

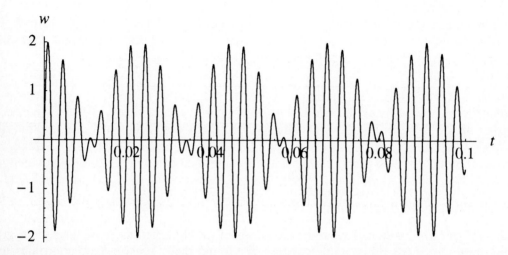

Next let's hear them. It is easier to hear the beats if the frequencies are closer, so change **w2** to have a frequency of 270 Hz before you play the waveform of the combination with Mathematica's **Play** command. What frequencies do you perceive when you play the sum?

In[11]:= **Play[w1 + w2, {t, 0, 5}];**

> Be patient. Mathematica takes a long time to play a waveform.

Problem 6. What is the lowest beat frequency that you can hear? That is, how close to 256 Hz can the frequency of **w2** be and you still detect the beats?

Problem 7. Suppose we add two waves, **w3** and **w4**, that have the same frequency, 256 Hz, but arrive at the same point ($x = 0$) out of phase. That is

$$w3 = \text{Sin}[2\pi ft], \quad w4 = \text{Sin}[2\pi ft + \phi] \qquad (5)$$

where, furthermore, the phase shift ϕ varies with time. In particular, choose $\phi = 2\pi\, 44\, t$. Here are the defining equations in Mathematica.

In[12]:= ϕ = 2 π 44 t;
w3 = Sin[2 π 256 t];
w4 = Sin[2 π 256 t + ϕ];

Plot the sum of these two waves from $t = 0$ to $t = 0.1$ s. Compare the plot you make with the the one we made. In particular, you might plot the sum of **w1** and **w2** and the sum of

w3 and **w4** on the same coordinate system. Try to write a sentence or two describing your results and then form a conclusion. You might also wish to **Play** the sum of **w3** and **w4**. (If you listen to AM radio or shortwave radio from distant stations at night, you notice fading, which is, in part, a result of the relative phase angle of several waves arriving at your antenna varying with time.)

■ 10.4 Interference: A Mathematica Approach

Interference refers to the addition of two or more waves with some definite phase relationship between the waves at some point in space. Beats are an interference phenomenon. We begin with the time variation of a wave arriving at a point, namely

$$E = A\sin(\omega t + \phi) \tag{6}$$

where E might symbolize the electric field intensity in a direction perpendicular to the direction of the motion of the wave, ω is the angular frequency of the disturbance, and ϕ is its phase angle relative to some more or less arbitrary zero phase angle. In what follows it will be convenient to take with no loss in generality both $\omega = 1$ and $A = 1$. (By now you surely have noticed how frequently we choose our parameters to be unity. It doesn't change the physics and it simplifies the mathematics. There is no harm in this especially if we are interested in concepts more than numbers.) In Mathematica we will express this wave and graph it for $\phi = 0$ as follows:

```
In[13]:= Clear["@"]
         e := Sin[t + ϕ]
         Plot[e /. ϕ -> 0, {t, 0, 4 π}, AxesLabel -> {"t", "E"}];
```

Problem 8. It is important to have an intuitive idea of what the phase shift ϕ does. Make a plot of a few waves with different phase shifts. You can do it this way:

```
In[16]:= Plot[Evaluate[Table[e, {ϕ, 0, π/4, π/16}]], {t, 0, 2 π},
         AxesLabel -> {"t", "E"}];
```

> The **Evaluate** command is a necessary but difficult to understand evil in an instruction such as this. Refer to *The Mathematica Book* if you desire understanding.

Does a positive phase shift ϕ move the sine wave to the left or the right?

Try making a similar graph but with negative phase shifts.

We will always be adding two or more waves with specific fixed phase shifts ϕ relative to each other. We will use Mathematica's **Sum** (or $\sum_{\square=\square}^{\square} \square$ from the palette) function to add the waves. You will see how it works by trying a few simple cases. After we **Clear**, here is the command that will add our waves.

In[17]:= **Clear["@"]**

$$e[n\_] := \sum_{i=0}^{n-1} (a\ \text{Sin}[\omega t + i \phi])$$

Problem 9. To understand what waves we are combining with the preceding function, in sequence find: **e[1]**, **e[2]**, **e[3]**, **e[4]**. Of what significance, if any, is **e[1]** to us?

10.4.1 Computing Intensity

In the case of light waves, for example, we do not perceive the time varying field intensity, which is quite different than watching water waves or a wave traveling down a string. In the case of electromagnetic waves what we perceive is the time average of the square of the electric field strength, or something proportional to that time average. Because the integrations can become quite challenging, Mathematica can be a big help in finding the time average. It's best to begin simple: Suppose we have just one wave. Then the square of the field strength is

In[19]:= **e[1]^2**

Integrating this is not too challenging, but combining two waves complicates things considerably. We generate the sum of two waves with our **e[2]**, then square it

In[20]:= **e[2]^2**

That doesn't look too terrible until you expand it.

In[21]:= **Expand[e[2]^2]**

Some hard-wrung and painful experience shows that we can save Mathematica and ourselves quite a bit of trouble if, before we integrate to find the time average, we use a command called **TrigExpand**.

In[22]:= **TrigExpand[e[2]^2]**

Although this result looks much worse than the simple **Expand[e[2]^2]**, there is one significant simplification. See if you can discover it.

We are ready for the time average, but first let's set

In[23]:= ω = **a** = 1;

10.4 Interference: A Mathematica Approach

Here is the time-average integral,

In[24]:= **intensity = $\dfrac{1}{2\pi}\displaystyle\int_0^{2\pi}$ TrigExpand[e[2]^2] dt**

which we must graph if we wish to interpret the results:

In[25]:= **Plot[intensity, {ϕ, -4π, 4π}, AxesLabel –> {"ϕ", "I"}];**

We obtain the classic two-source or double-slit interference pattern.

Problem 10. Compute and plot the intensity of the time average of a three-wave sum as a function of ϕ. Here is the expression you need to find the intensity.

In[26]:= **intensity = $\dfrac{1}{2\pi}\displaystyle\int_0^{2\pi}$ TrigExpand[e[3]^2] dt**

Problem 11. Compute and plot the intensity of the time average of several more sums, such as a four-wave sum and a five-wave sum. In each case graph the intensity as a function of the phase angle. What happens to the peak intensity as the number of elements in the sum increases? Can you find a nice simple proportion between the peak intensity and the number of sources? What happens to the width of the intensity maxima as the number of elements increases?

10.4.2 The *n*-Source Case

Can we find a general expression for the intensity in the case where we have n waves with successive phase differences ϕ? Let's try.

In[27]:= **intensity = $\dfrac{1}{2\pi}\displaystyle\int_0^{2\pi}$ TrigExpand[e[n]^2] dt**

This looks different than the textbook version of the same result obtained by combining phasors. Sometimes it pays to try some tricks. Since we see a cosecant squared factor in both terms of the numerator, let's try to factor it out with **TrigFactor**.

In[28]:= **nsource = TrigFactor[%]**

> The % sign means *previous result*. In this case we **TrigFactor** the previous result. Although sometimes the % symbol is a useful thing to use, we have found that using assigned names is better. Thus we could more wisely have found **TrigFactor[intensity]**.

You may recognize that this is the same as the textbook formula, which is usually written

$$I = I_0 \frac{\sin^2(n\phi/2)}{\sin^2(\phi/2)} \tag{7}$$

Problem 12. Graph this result as a function of ϕ using the rule **n -> 10**. Make the domain of the plot from -4π to $+4\pi$, and use the option **PlotRange -> All** to see all of the plot. Add axes labels.

If he or she hasn't already done so, at this point you should have your instructor shine a laser through diffraction gratings with several different slit spacings so you can relate what you see to our computations. You might also wish to think about the significance of the large number of radio astronomy telescopes in the Very Large Array (VLA) located near Socorro, New Mexico. This large observatory is not sending out waves, but rather it is receiving them from very distant galactic or extragalactic radio sources. Why do they use so many telescopes?

■ 10.5 Examples of Interference

What we have done so far is to find the amplitude when n waves of equal amplitude arrive at a point with successive phase differences of ϕ. One way in which two waves can interfere in this manner is shown in the following figure. Two sources that are oscillating in phase are separated by a vertical distance d. Waves from each of the sources head in a particular direction, θ, from the normal to the line connecting the sources. At some very distant point, or by means of a lens, these rays meet and interfere. The question is, how does the intensity vary with θ?

10.5 Examples of Interference

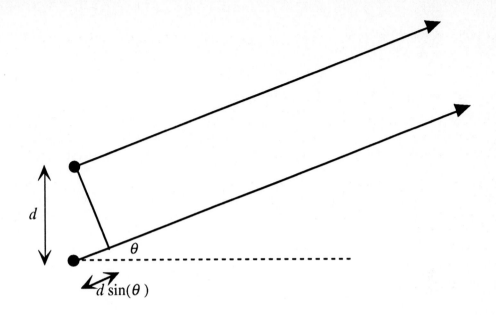

To begin, we see that the wave from the lower source must travel a longer distance, namely $d \sin(\theta)$, and that is what causes the phase difference. Your textbook will show that path difference and phase difference are related by the simple relationship

$$\frac{\phi}{2\pi} = \frac{d \sin(\theta)}{\lambda} \qquad (8)$$

Thus,

$$\phi = \frac{2\pi}{\lambda} d \sin(\theta) \qquad (9)$$

We pull together all of the important equations of the previous section into one cell and add another for Equation (9). Notice that we have combined several operations for the **intensity** and we have made it a function of *n*, the number of sources. This will better suit our purposes.

```
In[29]:= Clear["@"]
         a = ω = 1;

         e[n_] := Sum[(a Sin[ω t + i φ]), {i, 0, n-1}]

         intensity[n_] = TrigFactor[ 1/(2π) Integrate[TrigExpand[e[n]^2], {t, 0, 2π}] ];

         φ = (2π/λ) d Sin[θ];
```

Now we can find the intensity for any number of source as a function of the angle θ. For example, we can find

```
In[32]:= intensity[3]
```

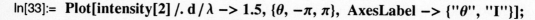

You can see that the intensity depends on the ratio d/λ, so we must specify this quantity before we graph the intensity. One choice for graphing the intensity is quite traditional: We show it below for a ratio of d/λ of 1.5.

```
In[33]:= Plot[intensity[2] /. d/λ -> 1.5, {θ, -π, π}, AxesLabel -> {"θ", "I"}];
```

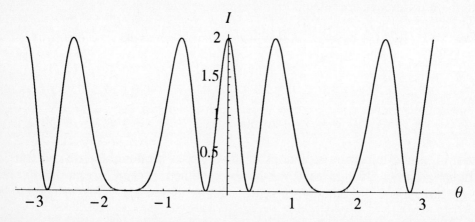

Another, perhaps better, choice is to make a polar coordinate plot, an idea that comes from Mechtly and Bartlett (1994). This gives a better feeling for how the intensity changes in various directions. Here is an example for two sources on the y-axis separated by a vertical distance d.

```
In[34]:= ParametricPlot[{intensity[2] Cos[θ], intensity[2] Sin[θ]} /. d/λ -> 1.5,
         {θ, -π, π}, AspectRatio -> Automatic, Compiled -> False,
         AxesLabel -> {"x", "y"}];
```

10.5 Examples of Interference

> The **Compiled->False** command is another necessary evil to avoid error messages.

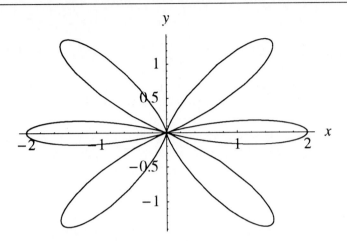

Note that in this plot *the intensity in a direction θ is proportional to the distance from the pole (origin) to the curve*. Also, keep in mind that we are showing the intensity variation in the *x-y* plane. Interference patterns are inherently three-dimensional.

Note that the polar plots will all have a *y*-axis symmetry, so we need only plot the right half if we keep the symmetry in mind.

In[36]:= **ParametricPlot[{intensity[2] Cos[θ], intensity[2] Sin[θ]} /. d / λ -> 1.5,
{θ, -π / 2, π / 2}, AspectRatio -> Automatic, Compiled -> False,
AxesLabel -> {"x", "y"}];**

Now we're going to do some tricky things so we can view several graphs at once. We begin by using the **Table** command to make a series of graphs with various d/λ ratios, in this case the ratio varies from 0.5 to 2.5 in steps of 0.5, exactly as in Mechtly and Bartlett (1994). To reduce clutter, we turn off the tick marks and the display of the graphs until we show the entire array. The options that do this are **Ticks -> False** and **DisplayFunction -> Identity**, respectively. Following that we put the graphs in an array with **GraphicsArray** and show them with **Show**. After you have made the graphics array, you may wish to select the graph and size it so it fits nicely on the screen or paper.

In[37]:= **Show[GraphicsArray[{Table[ParametricPlot[
{intensity[2] Cos[θ], intensity[2] Sin[θ]} /. d / λ -> r,
{θ, -π / 2, π / 2}, PlotRange -> All,
Compiled -> False, Ticks -> False, AxesLabel -> {"x", "y"},
DisplayFunction -> Identity], {r, 0.5, 2.5, .5}]}]];**

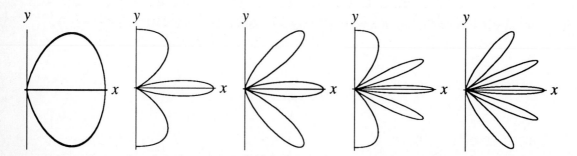

Problem 13. Describe how the intensity pattern changes as the distance between the sources increase – that is, as the ratio d/λ increases. Predict what will happen if this ratio is 3. Use one or more of the commands above to check your prediction. If you wish, make the step size smaller and make more graphs to see how the intensity patterns change as the ratio d/λ changes.

Problem 14. Make additional graphics arrays for d/λ varying from 0.5 to 2.5 in steps of 0.5 for a three-source array and a four-source array. For a given d/λ, what is the effect on the intensity pattern as the number of sources increases?

Consider a line of n sources confined to a specific distance w. To form a picture in your mind, we might put the sources along the y-axis with the first source at the origin and the nth source at $y = w$ (see the figure below). From the perspective of optics, we imagine a slit of width w through which the light passes.

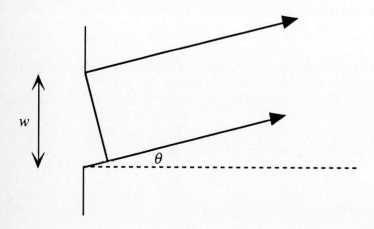

Suppose the phase angle between successive sources is ϕ, so the phase difference between the first source and the last source is $\varphi = n\phi$. We keep adding more and more sources keeping φ constant, which means that ϕ gets smaller and smaller. Recall that for small ϕ,

10.5 Examples of Interference

$\sin\phi = \phi$. Furthermore, remember that the intensity of the maxima increases as n^2, but in a real slit of fixed-width w the amount of light that goes through is fixed. So, we scale down the total energy radiated by all the sources by $1/n^2$. With these two modifications Equation (7) becomes

$$I = \frac{\sin^2(\varphi/2)}{(\varphi/2)^2} \qquad (10)$$

Now we graph this expression:

In[38]:= Plot$\left[\frac{\text{Sin}[\frac{\varphi}{2}]^2}{(\frac{\varphi}{2})^2}, \{\varphi, -5\pi, 5\pi\}, \text{PlotRange} \to \text{All}, \text{AxesLabel} \to \{"\varphi", "I"\}\right]$;

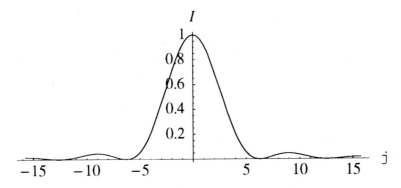

This effect is called *single-slit diffraction*. It is still an interference phenomenon, but it has been given this special name. In any real situation with multiple sources, such as a diffraction grating, we unavoidably get both inteference and diffraction. Each individual slit produces a pattern such as the one given above, while the several slits produce the interference patterns described earlier. We will discuss these results more extensively in the next section. Remember, φ is not the phase angle between successive sources within the slit, it is the phase angle between the waves coming from opposite ends of the slit. Thus in the case where parallel rays interfere, as in the case where the distance to the point of interference is large or when a lens is used to focus the rays, then

$$\frac{\varphi}{2\pi} = \frac{w\sin(\theta)}{\lambda} \qquad (11)$$

This is called Fraunhofer diffraction, as opposed to the case where the point of interference is much closer and we cannot make the assumption of parallel rays, which is called Fresnel diffraction.

Let's study Fraunhofer diffraction in the same way we did interference, namely through the use of polar plots.

In[39]:=
```
Clear["Global`*"]
φ := 2 π r Sin[θ];
Show[GraphicsArray[{Table[
```

$$\text{ParametricPlot}\left[\left\{\frac{\text{Sin}[\frac{\varphi}{2}]^2}{(\frac{\varphi}{2})^2}\text{ Cos}[\theta],\ \frac{\text{Sin}[\frac{\varphi}{2}]^2}{(\frac{\varphi}{2})^2}\text{ Sin}[\theta]\right\},\right.$$

{θ, −π/2, π/2}, PlotPoints −> 100,
Compiled −> False, Ticks −> False, AxesLabel −> {"x", "y"},
DisplayFunction −> Identity], {r, 0.1, 2, .4}]}]];

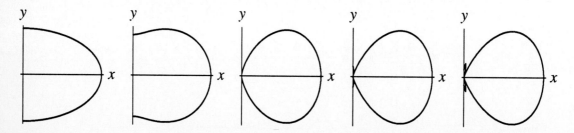

If you want to see one of these in more detail, you can choose **r** and then plot one of them, as follows:

In[41]:= r = 3;

$$\text{ParametricPlot}\left[\left\{\frac{\text{Sin}[\frac{\varphi}{2}]^2}{(\frac{\varphi}{2})^2}\text{ Cos}[\theta],\ \frac{\text{Sin}[\frac{\varphi}{2}]^2}{(\frac{\varphi}{2})^2}\text{ Sin}[\theta]\right\},\ \{\theta,\ -\pi/2,\ \pi/2\},\right.$$

PlotPoints −> 100, Compiled −> False, AxesLabel −> {"x", "y"}];

Now you can begin to see the suggestion of interference fringes in the little sidelobes.

Problem 15. Suppose the ratio of the slit width to λ is 0.25 as shown in the first frame, above. Also assume that we are shining light through such a slit onto a screen. What would you see?

Problem 16. What do you conclude happens to the diffraction pattern as the slit width gets bigger – that is, as the ratio $r = w/\lambda$ increases? You may wish to continue the set of graphs to larger values of r, say $r = 4$ or 5.

Problem 17. What would you see on a screen if the situation is as shown in the last frame in the figure above?

10.6 Diffraction and Interference

Because of the finite size of the individual slits in a diffraction grating, for example, in any real situation of interference we see both interference phenomena combined with single-slit diffraction. Thus, we must combine Equations 7, 9, 10, and 11 as follows:

$$I = \frac{1}{n^2} \frac{\sin^2(n\phi)}{\phi^2} \frac{\sin^2(\varphi/2)}{(\varphi/2)^2} \qquad (12)$$

where

$$\varphi = 2\pi \frac{w}{\lambda} \sin(\theta), \quad \phi = 2\pi \frac{d}{\lambda} \sin(\theta) \qquad (13)$$

We have divided the right-hand side by n^2 to normalize the intensity at $\theta = 0$ to unity. Making the substitutions gives a rather dramatic expression, namely

$$I = \frac{1}{n^2} \frac{\sin^2(2\pi n \frac{d}{\lambda} \sin(\theta))}{\sin^2(2\pi \frac{d}{\lambda} \sin(\theta))} \frac{\sin^2(2\pi \frac{w}{\lambda} \sin(\theta)/2)}{((2\pi \frac{w}{\lambda} \sin(\theta)/2))^2} \qquad (14)$$

What we are interested in understanding is the effect of varying the slit width w, the slit spacing d, the number of slits n, the wavelength λ, and the viewing angle θ. We write the expression for the intensity in Mathematica as a function of several variables so we can change these at will. Note that if we choose $d = w$ we simply have a single slit of width $2w$. Here is the expression for the intensity in Mathematica.

In[42]:= **int[λ_, n_, w_, d_, θ_]** := $\dfrac{1}{n^2} \dfrac{\text{Sin}\left[n\, 2\pi \frac{d}{\lambda} \text{Sin}[\theta]\right]^2}{\text{Sin}\left[2\pi \frac{d}{\lambda} \text{Sin}[\theta]\right]^2} \dfrac{(\text{Sin}[2\pi \frac{w}{\lambda} \text{Sin}[\theta]/2])^2}{(2\pi \frac{w}{\lambda} \text{Sin}[\theta]/2)^2}$

Now we can study the combination of interference and diffraction, which is always what we get because we cannot get one without the other. Once again, we must choose some parameters and vary others to get an idea of what happens when coherent light of a particular wavelength passes through a series of n slits with a spacing d and a width w. To begin we choose $\lambda = 0.1$, $n = 2$, $w = 1$, and $d = 3$ and ask how the intensity, normalized to be equal to 1 at $\theta = 0$, varies with the angle θ (refer again to the drawings to see the meaning of θ). The following plot illustrates what happens.

In[43]:= **Plot[int[.1, 2, 1, 3, θ], {θ, −π/16, π/16}, PlotRange −> All, PlotPoints −> 50, AxesLabel −> {"θ", "I"}];**

In this plot we see the interference fringes modulated by the diffraction pattern. We find 11 interference maxima in the central (zero-order) diffraction maximum. The first-order diffraction maxima can also be seen. Note that without the effects of diffraction, all of the interference maxima would have an intensity of 1. If you want to see the smaller interference lobes, change the **PlotRange** option to **PlotRange->{0, 0.2}**.

Problem 18. Use the results we have just obtained to study the effect of changing the wavelength on the diffraction/interference pattern. Start with $\lambda = 4$ and reduce it in steps until you can describe what happens to both the diffraction and the interference patterns.

With several parameters to change – for example λ, n, w, and d – it is sometimes useful to use three-dimensional plots to see the effect of changing the parameters on the interference/diffraction patterns (Mechtly and Bartlett, 1994). To illustrate, we use Mathematica's **Plot3D** command to graph the intensity as a function of θ and d.

In[44]:= **Plot3D[int[.5, 2, 1, d, θ], {θ, .00001, π/16}, {d, 1, 5}, PlotPoints –> {50, 75},**
   Shading –> False, MeshStyle –> {Thickness[.0001]}, PlotRange –> All,
   AxesLabel –> {"θ", "d", "I"}];

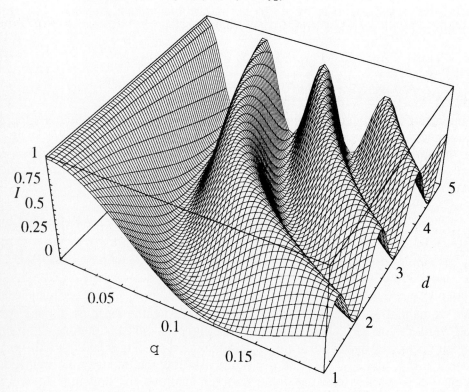

> To make the **Plot3D** command work without error messages, we avoid making θ identically equal to zero because the formula for the intensity becomes undefined at $\theta = 0$ because the intensity has terms of the form $\sin(\theta)/\theta$ However, since $\lim_{\theta \to 0} \sin(\theta)/\theta = 1$, we know the intensity at $\theta = 0$ is 1.

In this plot we have chosen to illustrate the two-slit case with $\lambda = 0.5$ and $w = 1$. We are graphing the intensity in the direction θ where θ is varying from zero to $\pi/16$ for values of slit separation that vary from 1 to 5. When $d = 1$, the slits merge into a single slit and we get a single slit diffrraction patter (front of the three-dimensional plot). As the slits separate, we get interference, and more and more interference fringes crowd into the central diffraction maxima.

Problem 19. Modify the **Plot3D** instruction to use θ and λ as variables. Graph λ from 0.5 to 2. Explain and interpret your results.

Problem 20. Experiment with different values for n. How does n effect the appearance of the graph?

Problem 21. Make a plot similar to the one we have made, but with diffraction effects removed.

■ 10.7 The Wave Equation

The differential equation governing many forms of wave motion is

$$\frac{\partial^2 y}{\partial x^2} = \frac{1}{c^2} \frac{\partial^2 y}{\partial t^2} \tag{15}$$

where y is the disturbance, which might be the displacement of a molecule of a string or the E-field of an electromagnetic wave, and c is the velocity of the wave. In the last few sections our language reflected our concern with optics, but in this section we will think of waves on a string. We return almost to the beginning of the chapter. Here is our expression for a wave.

In[45]:= **Clear["Global`*"]**

$$y := A \, \text{Sin}\left[2\pi \frac{x}{\lambda} - \omega t\right]$$

Does it satisfy the wave equation? We calculate both second partial derivatives as follows:

In[47]:= $\partial_{x,x} y$
$\partial_{t,t} y$

The sine terms are identical. Let's see what c must be.

In[49]:= $\text{Solve}\left[\partial_{x,x}\, y == \dfrac{1}{c^2}\, \partial_{t,t}\, y,\, c\right]$

So the wave can be traveling to the left or the right with velocity

$$c = \dfrac{\lambda \omega}{2\pi} = \lambda f \qquad (16)$$

which is exactly what we learned in Section 10.2. It's nice to know that our efforts in this chapter have not been wasted with an incorrect solution to the wave equation.

Solving the wave equation is a bit tricky because we are dealing with a partial differential equation, and partial differential equations are generally difficult to solve. Also, expressing certain partial derivatives in Mathematica is subtle. In any case we will use **NDSolve** to solve for a wave on a string (Wolfram, 1996). Here is the command that solves the differential equation.

In[50]:= **Clear["@"]**

$\text{sol} = \text{Flatten}[\text{NDSolve}[\{\partial_{t,t}\, y[x, t] == \partial_{x,x}\, y[x, t],\, y[x, 0] == e^{-(x^2/2)},$
$\text{Derivative}[0, 1][y][x, 0] == 0,\, y[-10, t] == y[10, t]\},\, y,$
$\{x, -10, 10\},\, \{t, 0, 20\}]]$

> This command takes a considerable amount of memory. Less memory is required if you reduce the time interval over which you want the solution.

We need to explain some of the terms. Recall that when you solve a differential equation you are actually integrating, so you introduce constants of integration. These are determined by placing conditions on the situation. In this case we will have two sets of initial conditions. The first condition is that at $t = 0$ the string has the form of an exponential expression, namely

$$y(x, 0) = e^{-x^2/2} \qquad (17)$$

giving the string some starting values. If you want to see this initial string deflection, here it is:

In[52]:= $\text{Plot}\left[y = e^{-x^2/2},\, \{x, -10, 10\},\, \text{PlotRange} \rightarrow \text{All}\right];$

It's like putting a little pulse on the string and then letting it go.

10.7 The Wave Equation

The second initial condition is the tricky one; that is, **Derivative[0, 1][y][x,0] == 0**. The meaning of this is that the first derivative (velocity) of the string at $t = 0$ will be zero. You cannot just write this as $\partial_t y[x, 0] == 0$ because this command has the effect of first substituting zero for t and then finding the derivative, which we all know is zero. Try it!

In[53]:= ∂_t y[x, 0] == 0

So, unfortunately, we must use the very subtle initial condition **Derivative[0, 1][y][x,0] == 0**, which means find the first derivative of **y[x, t]** with respect to the second variable, **t**, and then plug **x** and **0** in for **x** and **t**, respectively.

The boundary condition, **y[−10,t] == y[10, t]**, demands that the ends of the string behave in a symmetrical manner without tying them down. In one of the problems we will see how tying the ends of the string down changes the solution.

Finally, as is always the case with **NDSolve**, the solution is an interpolating function; in this case a two-dimensional interpolating function, and we use **Plot3D** to see the answer.

In[54]:= **Plot3D[Evaluate[y[x, t] /. sol], {x, −10, 10},
    {t, 0, 20}, PlotPoints −> 40, AxesLabel −> {"x", "t", "y"}];**
(∗Color∗)

In[55]:= **Plot3D[Evaluate[y[x, t] /. sol], {x, −10, 10}, {t, 0, 20}, PlotPoints −> 40,
    Shading −> False, PlotRange −> All, AxesLabel −> {"x", "t", "y"}];**
(∗Black & White∗)

Problem 22. Interpret the **Plot3D** graph you obtain. Where on the graph do you find the shape of the string at $t = 0$? What happens to the pulse as time goes on? What happens to the pulse when it reaches the ends, does it just go away and disappear? What do the ends of the string do?

Problem 23. Solve the same system, but replace the boundary condition with **y[−10, t] == y[10,t]==0**. This ensures that the ends of the strings stay fixed. Note that you may get an error message because these boundary conditions are not quite consistent with the initial conditions, but the results you get will be OK. Plot your solution just as we did, then interpret your graph.

Mostly Mathematica

1. Find and graph the solution to the differential equation

$$\sqrt{1 + t^2}\, x'(t) + x(t) = t, \quad t > 0 \tag{18}$$

given that $x(0) = 0$. Find a numerical solution to the same differential equation on the interval $t \in [0, 1]$. Make a table with t in the first column, the analytic solution in the second column, and the numerical solution in the third column. What is the largest percent difference in the two solutions?

2. Find and graph the solution to the differential equation

$$k = v'(t) - \frac{v(t)}{c} \qquad (19)$$

given that $v(0) = 0$ and k and c are constants. Integrate the solution for $v(t)$ directly to find $x(t)$ given that $x(0) = 0$.

3. Given the function

$$f(x) = e^{-x} \qquad (20)$$

first **Clear**, then create the corresponding function in Mathematica. Here is a list we have generated.

In[56]:= **ourlist = Table[i, {i, 1, 10}]**

How can we map our function over each element in the list? Look up the **Map** function, then map our function over the list.

Use **ListPlot** to graph this list. Then map **Log** over the list and plot it again (**Log** maps over the list automatically; you don't need the **Map** command). Now assemble **ListPlot**, **Log**, and **Map** and **Table** into a single command. Can you generate the list, map the function over it, take the log of each element, and then graph it in one command?

Explorations

1. Try to solve the wave equation with a more realistic initial condition for a plucked or bowed string (Giordano, 1997). Try an initial condition on **y[x, 0]** in which y increases linearly with x from $x = 0$ to some point on the string, say $x = 2$, and then decreases linearly with x from $x = 2$ to the other end of the string. Also, use boundary conditions characteristic of strings on an instrument, namely the ends are tied down.

2. In our study of interference and diffraction we have felt quite free to choose convenient values for our parameters, such as $\lambda = 1$, $d = 1$, and so on. Try to find some real values from devices in the laboratory and see if you can reproduce our results using these real values. Be sure to convert all values to SI. For example, a grating might have a distance between slits corresponding to 6000 slits/cm with opaque spaces between the slits that are equal to the slit widths. A visible wavelength to try might be 590 nm.

3. We have considered that successive phase difference between the sources are caused by path differences. In many applications, additional artifical phase differences are added to the sources electronically or with delay lines of some kind. Experiment with a four-or-five source array and try a variety of artifical phase differences between successive sources.

10.7 The Wave Equation

You should be able to sweep the lobes in different directions.

4. Don't underestimate the difficulty of this long project, but try to extend our results for interference and/or diffraction to two dimensions. Begin with two point-sources on the x-axis, then extend the results to sources on both the x-axis and the y-axis.

5. Relate the work in this chapter to the National Radio Astronomy Observatory's VLA interferometer. You might find some information about this antenna on the Internet. How far apart are the antennas placed? Why did they choose a Y formation? What is the slit width for this system? What kind of angular resolution do they achieve with the VLA?

References

Giordano, N. J. *Computational Physics,* Upper Saddle River, N.J.: Prentice Hall, 1997, p. 136.

Mechtly, B. and Bartlett, A. A. Graphical representations of Fraunhofer interference and diffraction, *Am. J. Phys.* 62 (1994): 501.

Wolfram, S. *The Mathematica Book*, 3rd ed, Champaign, Illinois: Wolfram Media and Cambridge, UK: Cambridge University Press, 1996, p. 893.

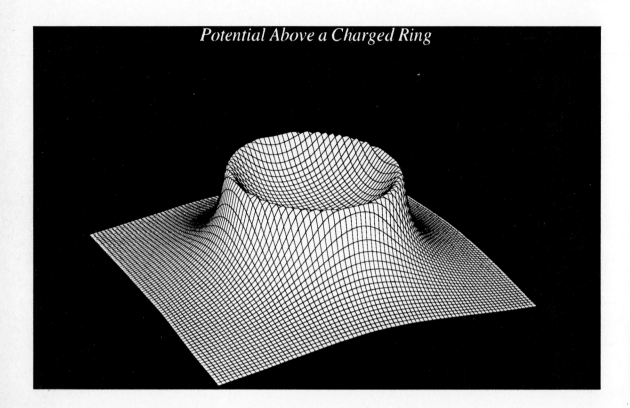

Potential Above a Charged Ring

CHAPTER 11 Electric Potential Problems

■ 11.1 Introduction

It might be wise to review the chapter in your textbook that deals with electric potential. In particular, we will make frequent use of the formula for the potential due to a point charge $+q$, namely,

$$V = k\frac{q}{r} = \frac{1}{4\pi\epsilon_o}\frac{q}{r} \qquad (1)$$

where k is Coulomb's constant. Recall that the components of the electric field are related to the potential by

$$E_x = -\frac{\partial V}{\partial x}, \quad E_y = -\frac{\partial V}{\partial y}, \quad E_z = -\frac{\partial V}{\partial z} \qquad (2)$$

and also recall that the force on a charge q is related to the electric field intensity **E** by the expression

$$\mathbf{F} = q\mathbf{E} \qquad (3)$$

and, finally, we will never forget that

$$\mathbf{F} = m\mathbf{a} = m\mathbf{v}'(t) \qquad (4)$$

There is a theorem in electrostatics, called Earnshaw's theorem, whose essence is that you cannot hold (trap) a charge in *stable* equilibrium with a static distribution of other charges. While you may be able to confine charges with magnetic fields, you cannot do that with a static group of charges. Here are some problems we will explore in pursuit of a deeper understanding of this theorem: (1) Can you trap a charge between two other stationary charges? (2) If a charge is constrained to move in a plane, can you hold a charge near the center of four charges at the corners of a square? (3) Can you trap a charge at the center of a cube whose eight corners each hold a stationary charge?

Problem 1. How does Earnshaw's theorem exclude the case of atoms – for example the hydrogen atom in which an electron is trapped by a proton?

■ 11.2 One-Dimensional Potentials

In[1]:= **Clear["Global`*"]**

We assume that in our laboratory charges are measured in bmoluocs, distances in retems, and potential in tlovs. Also in our laboratory we have found that a charge of 1 bmoluoc placed at a distance of 1 retem produces a potential of 1 tlov, making k unity, which is our reason for choosing these fictitious units. Furthermore, we will assume that $q = 1$ bmoluoc. To begin, let's attach this plus charge to the origin of our one-dimensional coordinate system, in which case the distance to the charge from an arbitrary point x on the coordinate systems is |x|. In Mathematica we write

In[2]:= $V[x\_] := \dfrac{1}{Abs[x]}$

We graph the potential with this command:

In[3]:= **Plot[V[x], {x, −1, 1}, AxesLabel −> {"x", "V"}, PlotRange −> {0, 20}];**

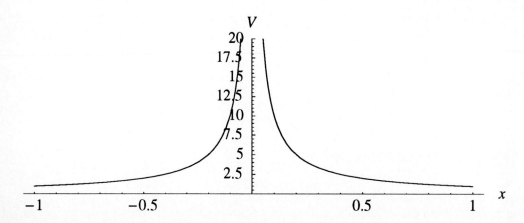

Problem 2. Is there any point x where you could put another charge, $+Q$, so that it would be in stable equilibrium? Explain.

We wish to plot the electric field intensity $E(x)$ for this case and it may be found by taking the derivative of the potential. However, Mathematica has trouble taking the derivative of |x|, so we use a simple theorem from algebra, namely:

$$|x| = \sqrt{x^2} \qquad (5)$$

Now we can rewrite the potential for our unit charge

In[4]:= **V[x_] :=** $\dfrac{1}{\sqrt{x^2}}$

We calculate the electric field with this command:

In[5]:= **e[x_] = $-\partial_x$ V[x]**

and graph it:

In[6]:= **Plot[e[x], {x, −1, 1}, PlotRange −> {−50, 50}, AxesLabel −> {"x", "E"}];**

Problem 3. Interpret this graph in terms of the force on a charge $+Q$ placed at some position x. Be sure to take into account the various signs of the force.

Next we consider the potential of two positive charges attached to the x-axis. Let's place a unit charge at $x = -1$ and the other unit charge at $x = 1$. Recalling that potential is a scalar quantity, we realize that we can simply add the potentials due to the individual charges.

Problem 4. Consider an arbitrary point with coordinate x on the x-axis. What is the distance between this point and the charge located at $x = -1$? What is the distance between this point and the charge located at $x = 1$? Express these distances in terms of square roots

11.2 One-Dimensional Potentials

rather than absolute values. Finally, write a formula for the potential at any point on the *x*-axis.

We write the potential for the situation described in Problem 4 in Mathematica.

In[7]:= $V[x\_] := \dfrac{1}{\sqrt{(x-1)^2}} + \dfrac{1}{\sqrt{(x+1)^2}}$

Now we plot the potential:

In[8]:= **Plot[V[x], {x, −2, 2}, AxesLabel −> {"x", "V"}, PlotRange −> {0, 20}];**

Problem 5. Is it possible to trap a charge $+Q$ between these two charges? Explain. Could you trap a charge $-Q$ between these two charges? What if the charge is allowed to move in another, perpendicular coordinate; can you trap a charge $+Q$ between two charges in that case?

We consider the motion of a unit charge $+Q$ between the two charges. The force on it is QE and if $Q = 1$, then the force is simple E. In Mathematica we write

In[9]:= $f[x\_] := -\partial_x \, V[x]$

We can solve the differential equation $F = ma$ with **NDSolve** as follows:

In[10]:= **solution = NDSolve[{f[x[t]] == v'[t], x'[t] == v[t], v[0] == 0, x[0] == 0.5}, {x, v}, {t, 0, 30}]**

Note that we started the particle at $x = 0.5$ with zero velocity and we are assuming a unit mass. We plot the solution.

In[11]:= **Plot[x[t] /. solution, {t, 0, 30}, AxesLabel −> {"t", "x"}];**

We can also make a phase-space plot of the motion of the charge.

In[12]:= **ParametricPlot[{x[t], v[t]} /. solution, {t, 0, 30}, AxesLabel −> {"x", "v"}, Compiled −> False];**

Problem 6. Describe the behavior of a charge $+q$ placed between two other $+Q$ charges, as described by the solution to the differential equation.

11.3 Electric Potential in Two Dimensions

You should have concluded from the work of the previous section that it is possible to trap a charge $+Q$ between two other plus charges, but only if the charge is constrained to move in one dimension. The world, however, is a three-dimensional world, and in this section we add the y-coordinate, working our way slowly to three dimensions. We assume, as before, that our two stationary unit plus charges are located at $(1, 0)$ and $(-1, 0)$.

Problem 7. Consider and arbitrary point (x, y) in the xy-plane. Write expressions for the distance between this point and each charge. Then write an expression for the potential at (x, y).

In Mathematica, the potential due to these two charges is

$$\text{In[13]:= } V[x\_, y\_] := \frac{1}{\sqrt{(x-1)^2 + y^2}} + \frac{1}{\sqrt{(x+1)^2 + y^2}}$$

In this case we use the **Plot3D** instruction to graph the potential as a function of both x and y.

```
In[14]:= Plot3D[V[x, y], {x, -4, 4}, {y, -4, 4}, PlotRange -> {0, 10},
        BoxRatios -> {1, 1, 2}, Shading -> False, PlotPoints -> 50,
        AxesLabel -> {"x", "y", "V"}, ViewPoint -> {0.726, -3.014, 1.356}];
(* Black and White*)

In[15]:= Plot3D[V[x, y], {x, -4, 4}, {y, -4, 4}, PlotRange -> {0, 10},
        BoxRatios -> {1, 1, 2}, PlotPoints -> 30, AxesLabel -> {"x", "y", "V"}];
(* Color*)
```

Another option for plotting surfaces such as this is the command **ParametricPlot3D**, which sometimes gives better graphs. With the next command you can see how to do this for the present problem.

```
In[16]:= ParametricPlot3D[{x, y, V[x, y]}, {x, -3, 3}, {y, -3, 3},
        PlotRange -> {0, 6}, PlotPoints -> 50, Boxed -> False,
        AxesLabel -> {"x", "y", "V"}, Shading -> False];
```

11.3 Electric Potential in Two Dimensions

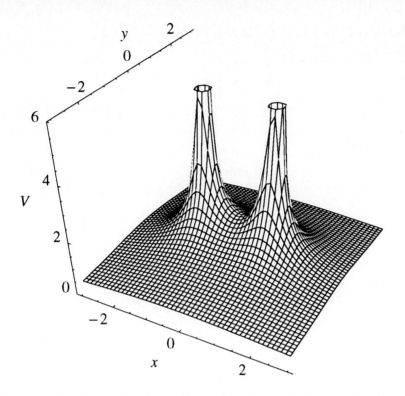

Problem 8. Study the three-dimensional graphs and decide if a charge $+Q$ can be trapped by two stationary plus charges. What will happen if the initial conditions for the $+Q$ are $x = 0$, $y = 0$, $\mathbf{v} = 0$? Is the particle in equilibrium at the point $(0, 0)$? Is the charge in *stable* equilibrium at $(0, 0)$? What will happen to the charge if the initial conditions are $x = 0$, $y = 0$, $v_x = 0$, $v_y = 0.001$? Can you have a charge $+Q$ in stable equilibrium in the vicinity of two stationary plus charges?

Clearly, with this distribution of charges, a charge $+Q$ located at the origin would tend to go down the potential hill in the y-direction. In our attempt to understand Earnshaw's theorem, we now locate two more unit plus charges to try to prevent motion in the y-coordinate. We put them at $(0, 1)$ and $(0, -1)$. Our thinking is that these two charges will tend to create a potential well in the y-direction. Our potential function now has four terms, one for each charge.

$$\text{In[17]:= } V[x\_, y\_] := \frac{1}{\sqrt{(x-1)^2 + y^2}} + \frac{1}{\sqrt{(x+1)^2 + y^2}} + \frac{1}{\sqrt{x^2 + (y-1)^2}} + \frac{1}{\sqrt{x^2 + (y+1)^2}}$$

We plot the potential with **ParametricPlot3D**.

In[18]:= **ParametricPlot3D[{x, y, V[x, y]}, {x, −2, 2}, {y, −2, 2},
AxesLabel −> {"x", "y", "V"}];**

Can you see if there is a small potential well near the origin?

Problem 9. Make a contour plot of the potential function. Use a **PlotPoints −> 100** option with it. Is there a potential minimum near the origin? Will a charge $+Q$ be in stable equilibrium at the origin?

Let's see if we can get a charge $+Q$ to move around in the potential well. First, we find the force per unit charge (the electric field) in the x-direction and make a function of it (see the equations in the Introduction).

In[19]:= **f[x_, y_] := −∂_x V[x, y]**

Likewise, we make a function of the y-component of the electrical force per unit charge.

In[20]:= **g[x_, y_] := −∂_y V[x, y]**

We clear some variables,

In[21]:= **Clear[x, vx, y, vy]**

then we solve the equation of motion in two coordinates with **NDSolve**, choosing some more or less arbitrary initial conditions and a unit mass.

In[22]:= **solution =
NDSolve[{f[x[t], y[t]] == vx'[t], x'[t] == vx[t], g[x[t], y[t]] == vy'[t],
y'[t] == vy[t], vx[0] == −.1, x[0] == 0, vy[0] == .1, y[0] == 0.1},
{x, vx, y, vy}, {t, 0, 50}]**

In[23]:= **ParametricPlot[{x[t], y[t]} /. solution, {t, 0, 50}, Compiled −> False,
AxesLabel −> {"x", "y"}];**

In this plot (on adjacent page) we see that the charge $+Q$ moves around in the potential well, but it does not leave it. A charge $+Q$ at the point (0, 0) with no velocity will remain at rest and is in stable equilibrium. Recall however, that we are implicitly constraining it to move in two dimensions only. What might it do in a third dimension?

11.3 Electric Potential in Two Dimensions

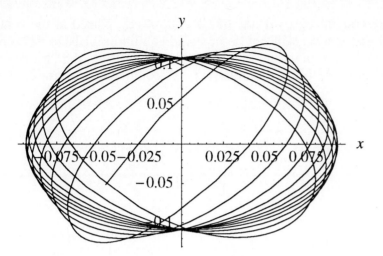

Problem 10. Is energy conserved for this motion? To answer this question, we begin by using the command **Flatten** to get rid of one set of set brackets.

In[24]:= **sol = Flatten[solution]**

The potential energy may be found this way:

In[25]:= **pe[t_] := u[x[t], y[t]] /. sol**

Create a function for the kinetic energy using **vx[t]** and **vy[t]**, then make a graph of the total energy as a function of time for the initial conditions shown above.

Problem 11. In two dimensions we have found a potential well in which we can trap a charge $+Q$ and, as a matter of fact, put it in stable equilibrium. Can you find a stable orbit of a charge $-Q$ around this group of four plus charges charges? This might not be easy.

Problem 12. Once again, the world is really three-dimensional, and we must take this third dimension into account before we decide if a charge $+Q$ is in stable equilibrium in the vicinity of these four stationary plus charges. Consider the same set of unit plus charges located at (1, 0, 0), (−1, 0, 0), (0, 1, 0), and (0, −1, 0). Pick an arbitrary point (x, y, z) and write an expression for the distance from the point to each of the charges. Then express the potential as a function of the three variables. Directly below you will find the answer , but don't cheat and look at it before giving an honest try.

In[26]:= $V[x\_, y\_, z\_] := \dfrac{1}{\sqrt{(x-1)^2 + y^2 + z^2}} + \dfrac{1}{\sqrt{(x+1)^2 + y^2 + z^2}} + \dfrac{1}{\sqrt{x^2 + (y-1)^2 + z^2}} + \dfrac{1}{\sqrt{x^2 + (y+1)^2 + z^2}}$

Problem 13. Produce some convincing evidence that a charge $+Q$ placed at the center (the origin) of this potential distribution will *not* be in stable equilibrium. Here are some possibilities to try: Make and interpret a graph of **V[0, 0, z]** as a function of *z*. Make and interpret a contour map of **V[0, y, z]**. Make and interpret a three-dimensional graph of **V[x, 0, z]**.

■ 11.4 Three-Dimensional Potential Functions

In[27]:= **Clear["Global`*"]**

We have failed to find a two-dimensional distribution of charges that will hold another charge of the same sign in stable equilibrium; the charge can always move perpendicular to the plane. In retrospect, this is probably not surprising. We now proceed to study a three dimensional distribution of charges and ask ourselves if it may be used to trap another charge. For the sake of symmetry we place unit plus charges at the eight vertices of a cube of side-length 2 centered at the origin. Here is the potential function.

In[28]:= **V[x_, y_, z_] :=**

$$\frac{1}{\sqrt{(x-1)^2 + (y-1)^2 + (z-1)^2}} + \frac{1}{\sqrt{(x-1)^2 + (y-1)^2 + (z+1)^2}} +$$

$$\frac{1}{\sqrt{(x-1)^2 + (y+1)^2 + (z-1)^2}} + \frac{1}{\sqrt{(x-1)^2 + (y+1)^2 + (z+1)^2}} +$$

$$\frac{1}{\sqrt{(x+1)^2 + (y-1)^2 + (z-1)^2}} + \frac{1}{\sqrt{(x+1)^2 + (y-1)^2 + (z+1)^2}} +$$

$$\frac{1}{\sqrt{(x+1)^2 + (y+1)^2 + (z-1)^2}} + \frac{1}{\sqrt{(x+1)^2 + (y+1)^2 + (z+1)^2}}$$

Problem 14. The question boils down to whether or not there is a potential minimum somewhere in this distribution, a likely position for a potential minimum being the origin. Use two-dimensional plots (hold two variables constant) such as

In[29]:= **Plot[V[x, y, z] /. {y -> 0, z -> 0}, {x, -1, 1}];**

Plot3D, **ParametricPlot3D**, or **ContourPlot** with one variable held constant to decide if there is a potential minimum at the origin. If a charge $+Q$ is placed at the origin with a tiny x-velocity, what will happen to it?

Mathematica has a package that makes a contour plot of a function of three variables. Actually, it plots surfaces at which the function of three variables has a particular value. To use it, you must first execute this instruction:

> In[30]:= << Graphics`ContourPlot3D`

Here is the **ContourPlot3D** instruction applied to our problem. Note that we are plotting two surfaces, namely those for which the value of the potential function is 4.61 and 4.6175.

> In[31]:= ContourPlot3D[V[x, y, z],
>   {x, −.5, .5}, {y, −0.5, .5}, {z, −.5, .5}, Contours −> {4.6175, 4.61},
>   PlotPoints −> {5, 5}];

Do you understand this graph? It is not simple. Does it convince you that there is not a potential minimum at the origin?

■ 11.5 The Uniformly Charged Sphere

> In[32]:= Clear["Global`*"]

For a last look at the problem we have been considering, suppose we bury a charge $+Q$ inside a sphere of radius R that contains positive charge uniformly distributed inside it. Certainly our freedom-loving charge $+Q$ cannot leak out of this. To deal with this problem, we will use, without proof, the results for the electric field **E** external to the sphere, at the sphere's surface, and inside the sphere as obtained from Gauss's law, and which are usually found in a physics textbook. Here are the results for the radial fields:

$$E_r = k \frac{q}{r^2}, \quad r \geq R$$
$$E_r = k \frac{q}{R^3} r, \quad r < R \tag{6}$$

where q is the total charge inside the sphere. We also use the fact that

$$V_B - V_A = -\int_A^B E_r \, dr \tag{7}$$

Using Equations (6) and (7) and the fact that the potential at infinity is defined to be zero, we find

In[33]:= $\mathbf{V[r\_]} = -\int_{\infty}^{r} \mathbf{k} \, \frac{\mathbf{q}}{\mathbf{r^2}} \, d\mathbf{r}$

which gives us the result we should expect. Furthermore, we can find the potential at $r = R$:

In[34]:= $\mathbf{V[R]}$

Take r for some interior point of the sphere. Then

In[35]:= $\mathbf{V[r\_]} = -\int_{R}^{r} \mathbf{k} \, \frac{\mathbf{q}}{\mathbf{R^3}} \, \mathbf{r} \, d\mathbf{r} + \mathbf{u[R]}$

Now we have a problem because we have two expressions for **V[r]**, one for the exterior and the surface and the other for the interior. We can solve the problem this way.

In[36]:= $\mathbf{newV[r\_]} = \mathbf{If}\left[r < R, \, \frac{\mathbf{k \, q}}{2 \, \mathbf{R}} \left(3 - \frac{r^2}{R^2}\right), \, \frac{\mathbf{k \, q}}{\mathbf{r}}\right]$

> If you remember neither how the **If** command works nor its syntax, type in a **??If** command.

Now we wish to plot this potential. To do this, we need values for the constants. Since, once again, we are interested in the behavior of the function rather than values at specific points, we will simplify matters by taking $k = 1$, $q = 1$ and $R = 1$. We define these as follows:

In[37]:= **rules = {k –> 1, q –> 1, R –> 1}**

and then we make the graph.

In[38]:= **Plot[newV[r] /. rules, {r, 0, 10}, AxesLabel –> {"r", "V"}];**

Clearly there is *not* a potential minimum at the center of the sphere, and a charge $+Q$ would not be in stable equilibrium there. What would happen to a charge $-Q$ in this sphere? Can charge ever be distributed uniformly in a sphere or any other volumn?

Problem 15. What is the variation of the electric potential with r inside a spherical surface of radius R with a charge q uniformly distributed over it? Use your textbook if necessary, and make a graph of the potential for values of r both inside and outside of the sphere. Can you trap a charge $+Q$ inside this sphere? What about a charge $-Q$?

Mostly Mathematica

1. Unit positive point charges are located in the xy plane at $(1, 0)$, $(0, 1)$, $(-1, 0)$, $(0, -1)$. Write an expression for the potential $V(x, y)$ at any point (x, y) in the plane, then graph this potential using the **Plot3D** command.

2. Plot the same potential as in the preceding problem using **ParametricPlot3D**.

3. A charge of +5 C is located at $x = 3$, and another charge of +3 C is located at $x = 5$. Where on the x-axis does the potential reach a local minimum?

Explorations

1. Given a distribution of point charges q_i located at (x_i, y_i), construct a Mathematica command or a series of Mathematica commands that calculates the electric potential V at any point (x, y) and calculates and plots the electric field intensity **E** at that point. Start simple with one charge, then proceed to a dipole and more complex charge distributions. Make use of Mathematica's $\sum_{\square=\square}^{\square} \square$ command.

2. Construct a model of scattering of alpha particles (Rutherford scattering) by the nuclei of atoms. You might wish to consult Gould and Tobochnik (1988) and De Jong (1991).

3. Explore the use of Laplace's equation to calculate electric potentials and fields. Excellent references include Gould and Tobochnik (1988) and Giordano (1997).

4. Use Mathematica to find and plot the potential, both off-axis and on-axis, of a uniformly charged ring and compare your results with the figure at the beginning of the chapter. Mathematica will be extremely helpful in doing the necessary integration. We also used cylindrical coordinates. Begin by studying the appropriate section in your textbook.

References

De Jong, M. L. *Introduction to Computational Physics*, Reading, Mass.: Addison-Wesley, 1991, p. 163.

Giordano, N. J. *Computational Physics*, Upper Saddle River, N.J.: Prentice Hall, 1997, p. 113.

Gould, H. and J. Tobochnik., *Computer Simulation Methods*, Reading, Mass.: Addison-Wesley, 1988, p. 223.

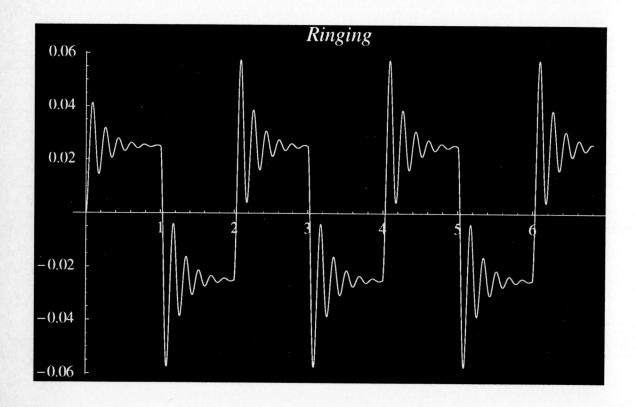

Chapter 12 Electrical Circuits

■ 12. 1 Introduction

In the first part of this chapter we will solve direct current (dc) circuits using Kirchhoff's laws: the junction rule and the loop rule. You may wish to review the appropriate section in your textbook. After that, we will analyze circuits with resistors, capacitors, and/or inductors for their transient behavior. These topics are traditionally found in various parts of a textbook: *RC* circuits in the section on capacitors, inductors and *LR* circuits following Faraday's law, and *LRC* circuits following closely thereafter. It will be helpful to remember that the voltage across a capacitor is

$$V_C = \frac{q}{C} \qquad (1)$$

while the voltage across an inductor is

$$V_L = -L\frac{di}{dt} \tag{2}$$

and, of course, the voltage across a resistor is *iR*. Finally, the current is

$$i = \frac{dq}{dt} \tag{3}$$

■ 12.2 dc Circuits: Kirchhoff's Law Problems

In[1]:= Clear["Global`*"]

A typical circuit is shown below and consists of resistors and capacitors. The first step is to label the various components and specify their values. The second step is to label currents in each branch and assign directions for the currents with the arrows. The direction does not have to be correct; if the current actually flows in the opposite direction, then the answer for the current will come out negative. Typically, but not always, the currents are the unknowns to be found.

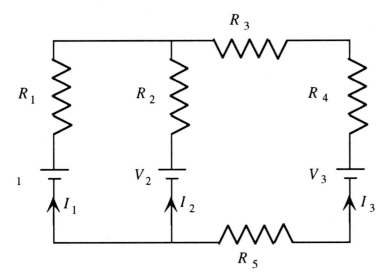

We will first apply the junction rule to the two junctions. Since all of the currents are flowing into the top junction and no currents are flowing out of the junction, we obtain (in Mathematica notation):

In[2]:= eqn1 = I1 + I2 + I3 == 0

Note that the junction on the bottom of the circuit involves no new currents; hence it does not give us an independent equation and we therefore don't use it.

We apply the loop rule to the left-hand loop starting at the bottom-left corner.

In[3]:= **eqn2 = V1 − I1 R1 + I2 R2 − V2 == 0**

The right-hand loop gives

In[4]:= **eqn3 = V2 − I2 R2 + I3 R3 + I3 R4 + V3 + I3 R5 == 0**

This gives us the three equations we need to solve for the three unknowns. Although we can get more equations, they will not help and might hinder. We could solve for the currents in terms of the other parameters symbolically, but we choose to assign some values.

In[5]:= **R1 = 1; R2 = 2; R3 = 2; R4 = 3; R5 = 1;**

In[6]:= **V1 = 12; V2 = 18; V3 = 10;**

In[7]:= **Solve[{eqn1, eqn2, eqn3}, {I1, I2, I3}]**

Problem 1. Find the three currents in the above circuit if all five resistors are 5 Ω and all three voltages are 10 V. You should obtain $I1 = I2 = \frac{4}{7}$ A, and $I3 = -\frac{8}{7}$ A.

Problem 2. Find a problem whose answer is in your textbook and solve it using these techniques.

■ 12.3 The *RC* Circuit

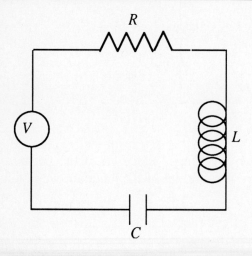

12.3 The RC Circuit

Above find an example of an *LRC* circuit with a driving voltage V. If the inductor is replaced with a wire, we have a *RC* circuit, while if the capacitor *C* is replaced by a wire, we have an *LR* circuit. We begin with *RC* circuits.

12.3.1 Charging and Discharging a Capacitor

In[8]:= **Clear["Global`*"]**

We apply Kirchhoff's loop rule to the *RC* circuit.

In[9]:= **eqn = R q'[t] + q[t]/c == V**

> We will use a small **c** because **C** is reserved.

You should realize that since the driving voltage V is not explicitly a function of time, Mathematica assumes it is a constant. Therefore, the circuit is essentially a resistor and a capacitor in series with a battery with voltage V. We can solve the equation with **DSolve**, choosing to start with the charge on the capacitor at zero.

In[10]:= **solution = DSolve[{eqn, q[0] == 0}, q[t], t]**

We can find the current by taking the derivative of the charge:

In[11]:= **i[t_] = ∂_t (q[t] /. solution)**

Now we must assign some parameters: Let's choose a capacitance of 1 μF, a resistance of 1000 Ω, and voltage of 10 V. Thus

In[12]:= **c = 1 10^{-6}; R = 1000; V = 10;**

Next, we plot the charge on the capacitor and the current (scaled down by 1000) as a function of time.

In[13]:= **Plot[{q[t] /. solution, i[t]/1000}, {t, 0, 5 10^{-3}}, AxesLabel -> {"t", "q/i"}];**

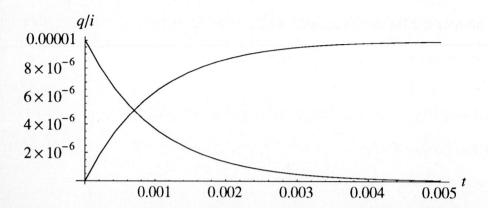

Problem 3. To consider a discharging capacitor, change the initial condition so that the initial charge is 10^{-5} C and set $V = 0$. Do this and make a graph of the charge on the capacitor as a function of time. Keep the parameters the same as above. Interpret the graph.

Problem 4. Instead of graphing $q(t)$, graph the natural log of $q(t)$ versus t. In Mathematica the natural log is **Log[x]**. What circuit parameters determine the slope of the straight line? Measure the slope from the printed graph and compare it with $-1/RC$.

12.3.2 Square-Wave Voltage Applied to an *RC* Circuit

In[14]:= **Clear["Global`*"]**

We wish to investigate the response of the *RC* circuit to a square wave. In other words, the source voltage V varies with time by flipping back and forth between 0 V and 1 V. We will use the following function to generate the square wave:

In[15]:= **V[t_] := If[Mod[t, 2] >= 1, 0, 1]**

which you can observe by graphing it.

In[16]:= **Plot[V[t], {t, 0, 10}];**

Applying Kirchhoff's law to the *RC* circuit gives the differential equation that governs the behavior of the circuit.

In[17]:= **eqn = R q'[t] + q[t]/c == V[t]**

We pick some simple values for the parameters:

In[18]:= **R = c = 1**

And, we use **NDSolve** to find a solution.

12.3 The RC Circuit

In[19]:= **solution = NDSolve[{eqn, q[0] == 0}, q, {t, 0, 10}]**

To find the current, we take the derivative of the charge on the capacitor.

In[20]:= **i[t_] = ∂_t (q[t] /. solution)**

Of course, we would like to see our results.

In[21]:= **Plot[{q[t] /. solution, i[t]}, {t, 0, 10}, AxesLabel -> {"t", "q/i"}];**

Problem 5. Use physical reasoning to identify which graph corresponds to the current and which to the charge on the capacitor.

Problem 6. Vary R and/or C until you can predict what happens with increasing and decreasing values of these parameters, then summarize your findings in a sentence or two.

Problem 7. Alternating current is sometimes converted by a half-wave rectifier to a direct current by applying a voltage from the half-wave rectifier of the form

In[22]:= **V[t_] := If[Sin[π t] > 0, Sin[π t], 0]**

to a RC circuit, which then smooths the voltage. Numerically solve the differential equation for the RC circuit given this voltage. Discuss how you might increase the effectiveness of the smoothing process; that is, how should the parameters be chosen to make voltage across the capacitor as steady as possible?

12.3.3 Sinusoidal Driving Voltages

In[23]:= **Clear["Global`*"]**

With a sinusoidal driving voltage, our differential equation for the RC circuit becomes

In[24]:= **eqn = R q'[t] + $\dfrac{q[t]}{c}$ == Sin[2 π f t]**

We will assume some nice easy values for the resistance and the capacitance, namely

In[25]:= **R = 1; c = 1;**

To prevent having to execute several instructions just to see a graph, we create a function that generates the graph with the execution of just one instruction, the function itself. Here is the function that solves the differential equation and plots the charge on the capacitor. Note that if we choose a unit capacitance, then the voltage across the capacitor is numerically identical to the charge.

```
g[f_, tmax_] :=
    Plot[Evaluate[q[t] /. NDSolve[{R q'[t] + q[t]/c == Sin[2 π f t], q[0] == 0},
        q, {t, 0, tmax}, MaxStepSize -> 1/(50 f), MaxSteps -> 2000]],
        {t, 0, tmax}, AxesLabel -> {"t", "V"}, PlotPoints -> 500];
```

> We found that for quite high frequencies the **MaxStepSize** option is important if you don't want to get nonsense from **NDSolve**. The reason is, of course, that as you increase the frequency, the rate of change of the various charges and voltages increases.

To obtain a graph, choose a frequency and the time duration of the solution, **tmax**. Here we choose **f = 10** and **tmax = 1**.

In[27]:= g[10, 1];

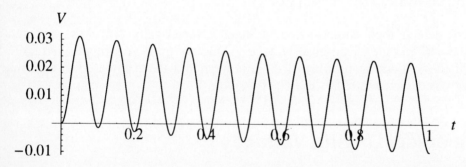

Problem 8. An *RC* circuit may be used as a low-pass filter. We imagine the input voltage to be the applied voltage, while the output voltage is taken across the capacitor. Use the function we have created to find the peak-to-peak value of the output voltage for a series of frequencies from 1 to 10. Then make a graph of the output voltage versus frequency. Describe what is meant by a low-pass filter.

■ 12.4 The *LRC* Circuit

As long as the driving voltage is constant, we can use **DSolve** to find the solution to the differential equation describing the *LRC* circuit. The differential equation itself is obtained by applying Kirchhoff's loop rule to the circuit. If the driving voltage varies in time, we can profitably turn to **NDSolve** for our solution. We begin with the simpler situation.

12.4.1 Simple Oscillations Without a Driving Voltage

In[28]:= **Clear["Global`*"]**

Here is the circuit equation obtained from Kirchhoff's loop rule.

In[29]:= **eqn := L q''[t] + R q'[t] + q[t]/c == 0**

We find the solution with **DSolve**:

In[30]:= **solution = DSolve[{eqn, q[0] == 1, q'[0] == 0}, q[t], t]**

Note that we have chosen to begin with an initial charge on the capacitor, but no initial current in the circuit. You are free to experiment with other initial conditions. We choose some simple values for the parameters so that we can see how the circuit behaves.

In[31]:= **R = .2; c = 1; L = 1;**

In[32]:= **i[t_] = ∂$_t$ (q[t] /. solution)**

Next, we plot the charge and the current as a function of time.

In[33]:= **Plot[{q[t] /. solution, i[t]}, {t, 0, 20}, AxesLabel -> {"t", "q/i"}];**

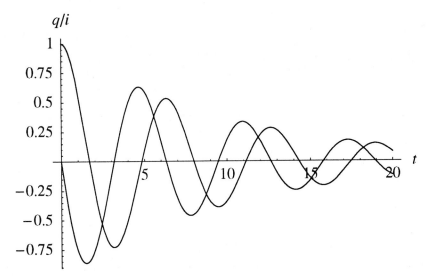

Problem 9. How do the oscillations change if you choose the resistance to be zero?

Problem 10. The frequency of the oscillations is predicted by theory to be $1/(2\pi\sqrt{LC})$. Use your graph in Problem 9 to verify that this is indeed the case when the resistance is zero.

Problem 11. This solution looks similar to the solution to the mass on a spring problem (with damping) that we obtained in Section 9.4. Make a table showing how the various parameters and variables are analogous. For example, the reciprocal of the capacitance C is the spring constant k, and the current is analogous to the velocity.

12.4.2 Driving the *RLC* Circuit with a Square Wave

In[34]:= **Clear["Global`*"]**

For our driving voltage we choose a square wave described by this function:

In[35]:= **If[Mod[t, 2] >= 1, −1, 1]**

You can graph the function to understand its behavior.

In[36]:= **Plot[If[Mod[t, 2] >= 1, −1, 1], {t, 0, 10}];**

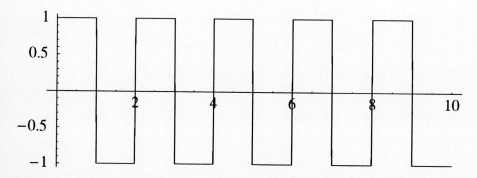

We choose some parameters that will illustrate the phenomena we wish to describe.

In[37]:= **R = 0.4; L = .02; c = .02;**

Here is the differential equation for the circuit.

In[38]:= **diffeqn = L q''[t] + R q'[t] + q[t]/c == If[Mod[t, 2] >= 1, −1, 1]**

And, here is the solution and a graph of the solution:

In[39]:= **solution = NDSolve[{diffeqn, q[0] == 0, q'[0] == 0}, q, {t, 0, 3.4}];**
        Plot[q[t] /. solution, {t, 0, 3.4}, AxesLabel −> {"t", "q"}, PlotPoints −> 100];

12.4 The LRC Circuit

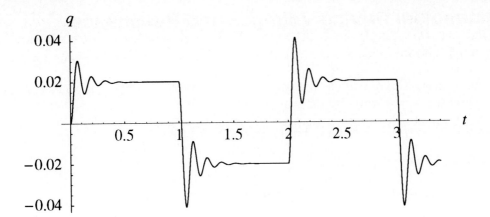

It appears that the circuit begins to oscillate at the beginning of each step, but the oscillations are quickly damped. This phenomenon is called *ringing*. In most cases it is undesirable. The figure at the head of the chapter is almost identical to this plot.

We can also find and graph the current in the circuit:

```
In[40]:= i[t_] = q'[t] /. solution;
         Plot[i[t], {t, 0, 3.4}, PlotRange -> All, AxesLabel -> {"t", "i"},
         PlotPoints -> 100];
```

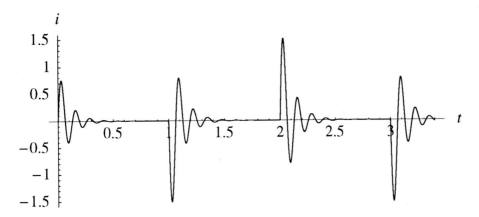

Problem 12. How can you reduce the ringing in the circuit? Try your ideas. Can you get it not to ring?

Problem 13. Try both some larger values for L and C and some smaller values for L and C, keeping the frequency of the driving force constant. How do the charge and current waveshapes change as L and C are changed?

12.4.3 Sinusoidal Driving Voltages and Resonance

In[41]:= **Clear["Global`*"]**

We replace the square wave with a sinusoidal wave whose amplitude is one volt, we choose some parameters, then we find the solution and graph it.

In[42]:= **diffeqn = L q''[t] + R q'[t] + q[t]/c == Sin[2 π f t]**

In[43]:= **R = 0.1; L = .02; c = .02; f = 2;**

In[44]:= **solution =
    NDSolve[{diffeqn, q[0] == 0, q'[0] == 0}, q, {t, 0, 6}, MaxSteps -> 4000];**

In[45]:= **Plot[q[t] /. solution, {t, 0, 6}, AxesLabel -> {"t", "q"}, PlotPoints -> 100];**

Problem 14. Make a graph of the current through the resistor as a function of time. Recall that the current is the derivative of the voltage. Refer to our previous work, if necessary.

We wish to investigate how the voltage on the capacitor varies with the frequency of the driving force. Because the voltage varies with time, this is a rather sophisticated problem. We could just plot the voltage as a function of time for a large number of frequencies and then measure its amplitude from these graphs, but we wish to let Mathematica do much of the work. We will use the root-mean-square (rms) voltage as a measure of its amplitude.

$$V_{rms} = \sqrt{\frac{1}{t_2 - t_1} \int_{t_1}^{t_2} (V(t))^2 \, dt} \qquad (4)$$

that is, the root of the mean of the square of the voltage as a function of time. Since we don't have a function that represents $V(t)$, only an interpolating function provided by Mathematica, we must do a numerical integration with **NIntegrate**.

To begin, you might wish to compare **NIntegrate** of a simple function with the results of **Integrate**. Let's try **Sin[x]** for our function. Here is **Integrate**:

In[46]:= $\int_0^\pi$ **Sin[x]** dx

and here is **NIntegrate**:

In[47]:= **NIntegrate[Sin[x], {x, 0, π}]**

What we need to do is solve the differential equation for some value of the frequency, then find the rms voltage over some time interval. There is some initial transient behavior, so

12.4 The LRC Circuit

we will solve the differential equation for $t = 0$ to $t = 3$, but we will only integrate from $t = 2.5$ to $t = 3$. The following function does this for some specified frequency.

In[48]:= **g[f_] := $\sqrt{\left(\dfrac{1}{.5}\text{ NIntegrate}[((q[s]/c) /. \text{Flatten}[}$**
    $\text{NDSolve}[\{L \ q''[t] + R \ q'[t] + q[t]/c == \text{Sin}[2\pi f \ t],$
    $q[0] == 0, q'[0] == 0\},$
    $q, \{t, 0, 3\}, \text{MaxSteps} \to 3000]])\verb|^|2, \{s, 2.5, 3.0\}]\right)}$

To see that this might take a long time to do, find the value of **g[f]** at some frequency, say $f = 2$, and time the computation with the **Timing** instruction, like this:

In[49]:= **Timing[g[2]]**

A 225-MHz computer took a little over a minute to do this one calculation. You can see that it might take quite a while to make a table of V_{rms} versus f, and even longer to make a more traditional graph. We choose to make a table and use **ListPlot**. We begin with the table

In[50]:= **resonance = Table[{f, g[f]}, {f, 6.9577, 8.9577, .1}]**

Out[50]= {{6.9577, 2.81038}, {7.0577, 3.05665},
    {7.1577, 3.35296}, {7.2577, 3.69714}, {7.3577, 4.09384},
    {7.4577, 4.63898}, {7.5577, 5.12845}, {7.6577, 5.87289}, {7.7577, 6.43229},
    {7.8577, 6.95816}, {7.9577, 7.06059}, {8.0577, 6.77888}, {8.1577, 6.17378},
    {8.2577, 5.4905}, {8.3577, 4.78646}, {8.4577, 4.2288}, {8.5577, 3.73026},
    {8.6577, 3.29721}, {8.7577, 2.98452}, {8.8577, 2.67152}, {8.9577, 2.44426}}

which took at least 30 minutes with a 225-MHz computer. You might wonder about the starting and ending values for the frequency in our table. The frequency at which the circuit naturally oscillates is

$$f = \dfrac{1}{2\pi\sqrt{LC}} \qquad (5)$$

and, for the parameters we have chosen, this frequency is 7.9577. So, we are attempting to center our graph on this frequency. Having calculated the table, we graph it with **ListPlot**. Notice that the voltage across the capacitor peaks when the frequency is near 7.9577. This phenomenon is called *resonance*, and 7.9577 is called the *resonance frequency*. Recall that we mentioned resonance in Chapter 9 in conjunction with driven oscillations of a mass on a spring.

In[51]:= **ListPlot[resonance, PlotRange -> All, AxesLabel -> {"f", "V$_{rms}$"}];**

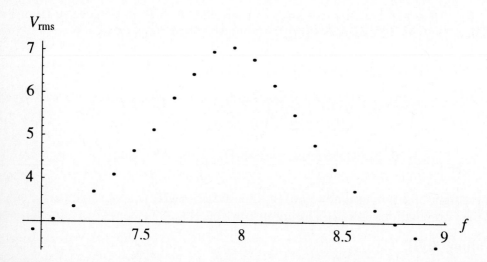

Problem 15. If you have a lot of computer time and patience, investigate the effect of the resistance in the circuit on the shape of the resonance curve. With this task you might cooperate with another group of students.

■ 12.5 The *LR* Circuit

In[52]:= **Clear["Global`*"]**

This one is all yours!

Problem 16. Draw a circuit diagram for an *LR* circuit.

Problem 17. Write the differential equation for the circuit with some driving voltage. Use the current $i(t)$ as the variable.

Problem 18. Express the differential equation in Mathematica syntax.

Problem 19. Solve the differential equation with **DSolve**, assuming the driving voltage is 0 and the inital current is 1. What is the solution?

Problem 20. Choose some simple parameter values, then graph the current as a function of time. Describe and explain your results.

Problem 21. Graph the voltage across the inductor as a function of time.

The stakes get higher.

Problem 22. Assume a triangle-wave voltage source. Here is one way to generate a triangle-wave voltage.

12.5 The LR Circuit

In[53]:= **V[t_] := If[Mod[t, 2] < 1, Mod[t, 1] − 0.5, −Mod[t, 1] + 0.5]**

In[54]:= **Plot[V[t], {t, 0, 5}];**

Choose $L = 0.001$ and $R = 10$ (this makes the inductive reactance very much less than the resistance, so the current in the circuit also has the form of a triangular-wave. Solve the differential equation over a time interval from 0 to 10 using **NDSolve**. Make a graph of **i[t]** versus **t**. It should be triangular.

Problem 23. The voltage across the inductor is $-L\, di/dt$. Find and graph this voltage. Explain your results.

Problem 24. Assume that of the total resistance of 10 Ω, 9Ω are actually in a resistor, while a resistance of 1 Ω is internal to the coil. This does not change the solution to the differential equation; it simply changes the voltage across the inductor to be $-(L\, di/dt + iR)$, where $R = 1\ \Omega$. This is the voltage you would actually observe if you put an oscilloscope across the inductor in this circuit. Plot this new voltage. It is relatively simple with today's function generators to set up an LR circuit to demonstrate this waveform. Perhaps you can persuade your instructor to do this. (See Duffy and Haber-Schaim, 1977).

Problem 25. Solve the differential equation for the situation when it is driven by a square-wave voltage source. Describe and explain your results. Can the current through an inductor ever take the form of a square wave?

Mostly Mathematica

1. Solve the system of equations: $2x + 3y + 4z = 10$, $x - y + z = 12$, and $2x - 3y + 12 = 0$.

2. Use **ParametricPlot** to make a series of plots of the parametric equations:

$$x = \sin(mt),\ y = \sin(nt) \qquad (6)$$

for several integer values of m and n.

3. Use calculus or the **FindMinimum** command to find the minimum of $y = \cosh(x) + \sinh(x)$ and the value of x that produces this minimum.

Explorations

1. Explore in more detail the use of RC and LR circuits in low- and high-pass filters. This should include making graphs of the output voltage of the filter as a function of frequency and also the gain of the circuit as a function of frequency. The output voltage may be taken across either the resistor or the capacitor (or inductor). Refer to you textbook for additional information.

2. Explore the use of an *LRC* circuit as a band-pass filter. The input voltage is across the series combination, while the output voltage is across one of the components, preferably the resistor. Begin by studying *LRC* circuits in your textbook.

3. Explore the behavior of a circuit in which R is connected in series with L and C in parallel. Solve its differential equation with a sinusoidal driving force and plot the voltage across the resistor as a function of time. Does this circuit have resonance properties?

4. A full-wave rectifier provides a voltage proportional to **Abs[Sin[2 π 60 t]]**, where 60 Hz is the line voltage frequency. In high-quality power supplies, this voltage is connected across an inductor L and a capacitor C in series. Of course, the inductor has some internal resistance R_L. The output voltage for the electronic circuit to be operated from this power supply is taken across the capacitor. In other words, there is a load resistance, R, across the capacitor. Investigate the capability of this circuit to provide a constant output voltage across the capacitor with different load resistances. Typical values might involve an inductor of several henries, a capacitance of tens of thousands of microfarads and a resistance of a few ohms.

Reference

Duffy, R. J. and U. Haber-Schaim, Establishing $V = L(di/dt)$ directly from experiment, *Am. J. Phys.*, 45 No. 2, February 1977: 170.

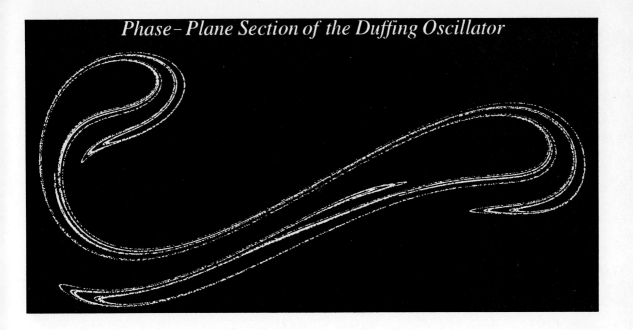

Phase–Plane Section of the Duffing Oscillator

CHAPTER 13 Return to Chaos

■ 13.1 Introduction

In Chapter 9, we studied various oscillating systems, one of which was the Duffing oscillator, and we found that under certain conditions, when the oscillator is both damped and driven, the solution may be chaotic. At that point *chaotic* meant there was no apparent pattern or periodicity in the solution. In this chapter we wish to examine this idea of chaos in more detail.

The study of chaotic systems began in earnest in the 1960s and saw enormous growth in the 1980s and 1990s. A simple but exciting account of many of these developments may be found in James Gleick's book *CHAOS Making a New Science* (1987). This chapter also represents an attempt to deal with a current topic in physics, a topic that became a major focus in the physics of the latter part of the 19th century.

> The bifurcation diagrams and the Poincaré maps take an enormous amount of memory. They also take a lot of time to print. If you plan to use this chapter in notebook form, you may wish to cut out these diagrams. An alternative is to cut the chapter in pieces and save it as several notebooks.

■ 13.2 The Quadratic Map: A Simple Approach

In[1]:= **Clear["@"]**

We begin with a problem from biology. A population of butterflies inhabits a certain isolated island. Each year the butterflies lay their eggs and die, and the following spring a new population hatches. How does the new population depend on the old population? It is conceivable that the more butterflies there were in year 0, the more there will be in year 1. Thus we might expect the number of butterflies, x, in year n to be proportional to the number in year $n-1$. But this leads to exponential growth and the butterflies would run out of food resources. In that case the population might decline. There must be another factor in the equation for butterfly population, one that decreases as the number of butterflies increases. Although it is doubtful that the situation is ever quite this simple, biologists have created the following equation to describe one possible scenerio:

$$x_{n+1} = 4\lambda x_n (1 - x_n) \qquad (1)$$

where λ is a parameter that depends on the particular circumstances on the island and the butterflies themselves. In this case x refers not to the actual number of butterflies but rather to a percentage or a fraction ($x < 1$) of some maximum number of butterflies, perhaps 100,000. With this restriction and an additional restriction on λ, namely $0 \leq \lambda \leq 1$, avoid the embarassing possibility of a negative x. With this background, we turn to a study of how the population of butterflies varies with time.

The so-called *difference equation*, Equation (1), an equation that, by the way, is quadratic in x, can be expressed in Mathematica in the following manner:

In[2]:= **x[n_] := 4 λ x[n – 1] (1 – x[n – 1])**

however, it is better to use a slightly different form

In[3]:= **x[n_] := x[n] = 4 λ x[n – 1] (1 – x[n – 1])**

in which case Mathematic remembers each of the **x[n]** as they are calculated. Since we are not biologists, we pick an arbitrary value for λ just to see how the population might change with from year to year. We will make a a couple of graphs, one starting with a small population, x near 0, and another with a large population, x near 1. We will model the situation corresponding to a large starting population.

13.2 The Quadratic Map: A Simple Approach

```
In[4]:= Clear["@"];
        λ = 0.7; x[0] = 0.9;
        x[n_] := x[n] = 4 λ x[n - 1] (1 - x[n - 1])
        ListPlot[
            Table[{n, x[n]}, {n, 0, 15}], PlotRange -> All, AxesOrigin -> {0, 0.3},
            AxesLabel -> {"n", "x_n"}, Prolog -> PointSize[.01]];
```

Our interpretation of this graph is that the population drops dramatically in year 1e, but then jumps back up again, oscillates back and forth for a number of years, but slowly settles down near $x = 0.6$.

Problem 1. Model the same situation except start with a small population such as $x[0] = 0.1$. Interpret your graph.

Problem 2. Model a situation with a new value for λ, $\lambda = 0.8$. Experiment with various starting values for x. Carry out the plot to values of n as large as 50. Interpret your graph. Is the population a periodic function? What is its period?

We repeat the experiment with $\lambda = 0.95$ and a starting value of 0.1.

```
In[6]:= Clear["@"];
        λ = 0.95; x[0] = 0.1;
        x[n_] := x[n] = 4 λ x[n - 1] (1 - x[n - 1])
        ListPlot[
            Table[{n, x[n]}, {n, 0, 100}], PlotRange -> All, AxesOrigin -> {0, 0.1},
            AxesLabel -> {"n", "x_n"}, Prolog -> PointSize[.005]];
```

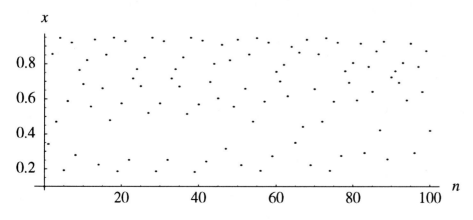

Is a pattern discernable in this graph? Is the population periodic? Could the population of butterflies be chaotic? Note that we have seen the population stabilize at some value, we have seen the population oscillate, and we have seen the population possibly turn chaotic.

To study this problem further, we will want more powerful techniques.

■ 13.3 The Quadratic Map: Part II

We begin by generating several of the population values in the same way as before

    In[8]:= **Clear["@"];**
            $\lambda = 0.7$; **x[0] = 0.9;**
            x[n_] := x[n] = 4 λ x[n − 1] (1 − x[n − 1])
            Table[x[n], {n, 0, 5}]

We now take a different approach to finding these points, which, by the way, are known as (part of) the *orbit* of $x = 0.9$. We define the function

    In[10]:= **f[x_] := 4 λ x (1 − x)**
             $\lambda = 0.7$;

In traditional mathematical notation we have defined

$$f(x) = 4\lambda x(1-x) \qquad (2)$$

and let our *difference equation* be defined by the expression

$$x_{n+1} = f(x_n) \qquad (3)$$

(A function is sometimes called a mapping, and we are dealing with a quadratic function, hence the name *quadratic map*.)

Now we apply Mathematica's **NestList** function. See for yourself what it does.

    In[12]:= **NestList[f, 0.9, 5]**

How does this list compare with the previous one? Try these operations:

    In[13]:= **f[.9]**
            f[f[.9]]
            f[f[f[.9]]]
            f[f[f[f[.9]]]]

The **NestList** function *iterates* $f(x)$ on whatever starting value is given. The successive values it obtains are, in fact, called the *iterates* of f. The list of successive iterates is called an *orbit* of the point $x = 0.9$. Clearly we can choose another starting point and find another orbit. Let's do that.

13.3 The Quadratic Map: Part II

    In[17]:= **NestList[f, 0, 5]**

In this case we have found a fixed point of f, namely zero. A *fixed point* is a point that is unchanged by additional applications of f.

Problem 3. Is $x = 1$ a fixed point of f? Is 0.58334 a fixed point of f for $\lambda = 0.6$?

Problem 4. Is $x = 0$ a fixed point of $x_{n+1} = \sin(x_n)$? You should not need Mathematica to answer this question.

If we experiment with various values of λ as follows:

    In[18]:= **λ = 0.8;**
             NestList[f, 0.8, 20]

we will find a new behavior of f. Note that the orbit very quickly becomes *periodic*. It is said to be *attracted* to an orbit or cycle of period 2. The *period* of the orbit is the least number of iterations that bring the iterates back to the starting point, hence, in this case, 2.

Problem 5. Try different starting values, x, for $\lambda = 0.8$. Does the orbit always become periodic for $0 < x < 1$? If it does, then $0 < x < 1$ is a *basin of attraction* for this orbit.

Notice from the orbit that the we did not immediately jump to this cycle of period 2. The orbit gradually moved into it. This initial behavior is called *transient* because it does not persist. After enough iterations the orbit becomes periodic; it is not periodic at the beginning of the orbit.

Clearly we should not go on hunting for various behaviors of the quadratic map, $f(x)$, by just trying various values of λ and different starting values for our iteration. We need to become more methodical and sophisticated, perhaps even elegant. Our plan is to make a graph of an orbit, after throwing out the transient behavior, for a large number of values of λ between 0 and 1.

We begin with another Mathematica function, **Nest**. **Nest** does not give all the iterates, only the last. Compare the output of

    In[19]:= **NestList[f, 0.8, 20]**

with the output of

    In[20]:= **Nest[f, 0.8, 20]**

which just gives us the 20th iterate, assuming $x = 0.8$ is called the zeroth iterate, $f(0.8)$ the first iterate, and so on. What does this function do?

    In[21]:= **NestList[f, Nest[f, 0.8, 5], 4]**

First it iterates five times discarding all but the fifth iterate, which then becomes the first iterate for **NestList**, which then provides the fifth iterate of **Nest** plus four more.

Problem 6. Explain what this function does. Experiment with it until you understand it. Write a one-sentence description of what it does.

In[22]:= **iterate[m_, n_] := NestList[f, Nest[f, 0.5, m], n]**

We will make use of another function to plot our points on the graph. Mathematica's **Point** function will be used. **Point[{r, s}]** plots a point at the coordinate (r, s). The next function, **h[y]**, takes a number y and creates a point with coordinates (λ, y).

In[23]:= **h[y_] := Point[{λ, y}]**

If we apply **h[y]** to the set of iterates we create a set of points with λ as abscissa and the iterate as ordinate. We can do this with the **Map** function.

In[24]:= **Map[h, iterate[4, 5]]**

Here is how to graph the iterates (we choose $\lambda = 0.95$):

In[25]:= **λ = 0.95**
 Show[Graphics[Map[h, iterate[5, 10]]]];

Now we must graph this for a series of λ values, not just one. We begin with a table of these points for various λ, but we immediately go to graphing the points

In[27]:= **Table[Map[h, iterate[5, 10]], {λ, 0, 1, .5}]**

In[28]:= **Show[Graphics[Table[Map[h, iterate[5, 10]], {λ, 0, 1, .1}]]];**

Now that we have the basics down, we proceed to add the extras

In[29]:= **Show[Graphics[Table[Map[h, iterate[5, 10]], {λ, 0, 1, .1}]],**
 AxesLabel -> {"λ", "x$_n$"}, Axes -> True, PlotRange -> {0, 1}];

To aid in our experimentation, we turn this into a function so we can vary our parameters more easily. It is important to realize that the functions f, **iterate**, and h must be defined for our **quadratic** function to work.

In[30]:= **quadratic::usage "quadratic[λstart_,λend_,mtrans_,nkeep_,ndiv_] makes a graph of the iterates of the quadratic map f[x] = 4λ x(1−x) for values of the parameter λ between λstart and λend. The number of transient iterates before graphing begin is mtrans, and the number of iterates graphed is nkeep +1. The x−axis interval is divided into ndiv parts."**

13.3 The Quadratic Map: Part II

```
quadratic[λstart_, λend_, mtrans_, nkeep_, ndiv_] :=
    Show[Graphics[{PointSize[0.001], Table[Map[h, iterate[mtrans, nkeep]],
        {λ, λstart, λend, (λend − λstart)/ndiv}]}],
    Axes −> True, AxesLabel −> {"λ", "x_n"},
    PlotRange −> {0, 1}];
```

We try a simple version to make sure everything is working (which takes almost no time).

In[32]:= **f[x_] := 4 λ x (1 − x);**
        iterate[m_, n_] := NestList[f, Nest[f, 0.5, m], n]
        h[y_] := Point[{λ, y}]
        quadratic[0, 1, 10, 10, 10];

Then we get really good (which takes time and lots of memory). We throw away the first 100 iterates at each value of λ and we graph the next 201 iterates for all admissible λ.

In[35]:= **quadratic[0, 1, 100, 200, 400];**

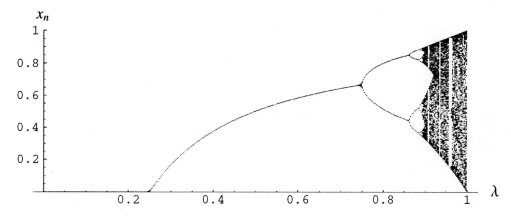

Now we are in a position to understand the behavior of this function and, therefore, the butterfly population for various values of λ, the *growth parameter*. Below λ ≃ 0.25, the population will always tend to zero. From λ ≃ 0.25 to almost 0.75, we have orbits of period 1. The population is stable at some number uniquely determined by λ. At λ ≃ 0.75, a *bifurcation* occurs and we have orbits of period 2. In fact, this graph is sometimes called a *bifurcation diagram*. At λ ≃ 0.86, another bifurcation occurs and the orbits have period 4. This means the population of butterflies will go through a cycle of a different population each year for 4 years, after which the population repeats the cycle. To see this period-doubling phenomenon in more detail, we graph the bifurcation diagram over a much smaller range of λ.

In[36]:= **quadratic[.85, .9, 100, 200, 100];**

Problem 7. Interpret this section of the bifurcation diagram. Are there other period doublings that occur?

Finally, we see that after a few period doublings the population becomes chaotic. The population never repeats from year to year. Do we actually have populations of insects that behave this way? I don't know the answer to that question. We do have periodic population variations; the 17-year locusts in my own yard this summer attest to this.

Problem 8. Using the **quadratic** function we created, expand the gap in the bifurcation diagram from $\lambda \approx 0.956$ to $\lambda \approx 0.965$. What is happening there? Do we ever get period 3 or period 5 orbits? To convince yourself of your conclusion, you might also set $\lambda = 0.958$ and try the **iterate** function again

In[37]:= λ = **0.958;**
                Show[Graphics[Map[h, iterate[100, 200]]]];

which makes a graph of 201 iterates at $\lambda = 0.958$.

■ 13.4 Return of the Duffing Oscillator

In[38]:= **Clear["Global`*"]**

In Chapter 9, we studied the Duffing oscillator characterized by a potential energy function

In[39]:= $U[x\_] = -\dfrac{x^2}{2} + \dfrac{x^4}{4}$

and a force per unit mass of

In[40]:= $f[x\_] = -\partial_x U[x]$

We can plot these functions as follows:

In[41]:= **Plot[{U[x], f[x]}, {x, −3, 3}, AxesLabel −> {"x", "U/F"}];**

We may think of the Duffing oscillator as a unit mass moving back and forth in a double potential well. (We choose the particle-in-a-well metaphor because it is simple. The Duffing oscillator need not apply to particles at all.)

13.4.1 Bifurcation Diagram for the Duffing Oscillator

In[42]:= **Clear["Global`*"]**

13.4 Return of the Duffing Oscillator

To make it behave in a chaotic fashion, we must add damping and driving forces to obtain the following equation of motion for the acceleration per unit mass from Newton's second law:

In[43]:= **a = v'[t] == x[t] – x[t]^3 – γ v[t] + d Cos[ω t]**

In Chapter 2, we showed how to use an equation like this to find the position and the velocity as a function of time. From Section 2.3.3, we obtain

In[44]:= **v[n_] := v[n] = v[n – 1] + a h**
x[n_] := x[n] = x[n – 1] + v[n] h

The point is that these equations are, exactly like the quadratic map, *difference* equations. A differential equation is solved numerically with difference equations. **NDSolve** works the same way, but it is much more sophisticated. Give it a particular *v* and *x* and it will find a new *v* and *x*. That is exactly how we will use it in this section to explore the chaotic behavior of the Duffing oscillator. Just as in the case of the quadratic map, we will use **NestList** and **Nest** to iterate, and just as in the case of the quadratic map, we will throw away many initial iterates to bypass the transient behavior. Instead of varying λ as we did in the quadratic map, we will vary the magnitude of the driving force per unit mass, **d**.

Here is an important but subtle point. Experience shows that a driven oscillator such as the Duffing oscillator typically settles down in some periodic orbit whose period is some multiple of the period of the driving force, $T = 1/2\pi\omega$. If $x(t)$ is in fact periodic with period T, then if we sample $x(t)$ once per driving force cycle, we will always get the same x with repeated samples.

Problem 9. Suppose $x(t)$ is periodic with a period $2T$, where T is the period of the driving force. Suppose we sample $x(t)$ once per driving force cycle. How many different values of $x(t)$ will we sample if we continue sampling in this way? How will this change if the period of oscillation is $3T$? What will happen to our set of samples if the motion is not periodic but chaotic?

If you understand the point just made, then we are ready to proceed to construct a bifurcation diagram for the Duffing oscillator. We assign values for the damping constant γ, the driving force d, and the driving frequency ω, and we calculate the period of the driving force:

In[44]:= **γ = 0.1; ω = 1.4; d = 0.20; T = $\frac{2\pi}{\omega}$;**

Given some initial conditions for *x* and *v*, **xstart** and **vstart**, respectively, the following application of the **NDSolve** command will find the values of *x* and *v* at the end of one period of the driving force.

$$\text{NDSolve}[\{v'[t] == x[t] - x[t]^3 - \gamma\, v[t] + d \cos[\omega\, t],$$
$$x'[t] == v[t],\ x[0] == \text{xstart},\ v[0] == \text{vstart}\},\ \{x, v\},\ \{t, 0, T\}]$$

However, we want to apply **NDSolve** repeatedly, each time beginning with the previous values of x and v. We create the following function to return new values for x and v after each cycle:

In[45]:= $g[\{\text{xnew}\_,\ \text{vnew}\_\}] :=$
$\{x[T], v[T]\}\ /.\ \text{Flatten}[\text{NDSolve}[\{v'[t] == x[t] - x[t]^3 - \gamma\, v[t] + d \cos[\omega\, t],$
$x'[t] == v[t],\ x[0] == \text{xnew},\ v[0] == \text{vnew}\},\ \{x, v\},\ \{t, 0, T\}]]$

Then

In[46]:= **Nest[g, {0, 0}, 10]**

will produce the last of 11 iterates of x and v starting with $x = v = 0$, counting (0, 0) as the first iterate. Also

In[47]:= **NestList[g, Nest[g, {0, 0}, 10], 15]**

will produce 16 iterates of x and v starting with the last iterate of **Nest**.

Now we need a function that takes the x-values of this list and makes it the y-coordinate of a point whose first coordinate is the driving parameter d. Here it is.

In[48]:= $f[\{x\_,\ y\_\}] := \{d, x\}$

We will **Map** this function over the iterates to get a set of points to plot.

In[49]:= **Map[f, NestList[g, Nest[g, {0, 0}, 10], 15]]**

This gives us a graph for only one value of the driving parameter d; to get the entire bifurcation diagram, we need to vary d. We can do this with a **Table** command.

In[50]:= **Flatten[Table[Map[f, NestList[g, Nest[g, {0, 0}, 10], 15]],**
    {d, .2, .4, .05}], 1]

Notice that to **ListPlot** this table we must first get rid of the outside pair of set brackets. We do this with **Flatten[list, 1]**. Now we are ready to plot a sample graph.

In[51]:= **ListPlot[Flatten[**
    Table[Map[f, NestList[g, Nest[g, {0, 0}, 10], 15]], {d, .2, .4, .05}], 1]];

This graph is not very meaningful. Part of the problem is that we are not throwing away enough iterates to get rid of the transient behavior. Also, it would be helpful if we had a

13.4 Return of the Duffing Oscillator

more sophisticated approach, which we now develop with a new function, **bifurcation**. This function needs the **f** and **g** functions, defined earlier, to be in memory before it works.

In[52]:= **bifurcation::usage** "bifurcation[dstart, dend, mtrans, nkeep, ndiv] makes a graph of the iterates of the solution to the differential equation of the Duffing oscillator for values of the driving paramter between dstart and dend with ndiv divisions of this interval. The number of transient iterates discarded is mtrans, the number plotted is nkeep + 1."

bifurcation[dstart_, dend_, mtrans_, nkeep_, ndiv_] :=
ListPlot[Flatten[Table[
Map[f, NestList[g, Nest[g, {0, 0}, mtrans], nkeep]],
{d, dstart, dend, (dend − dstart) / ndiv}], 1],
AxesOrigin −> {dstart, −1.6}, PlotStyle −> PointSize[0.001],
AxesLabel −> {"Driving Force", "x(nT)"}];

With several hours, to spend we can make an awesome bifurcation diagram. It is best to start with a small number of points and proceed to a larger number.

In[54]:= $\gamma = 0.1;\ \omega = 1.4;\ d = 0.20;\ T = \dfrac{2\pi}{\omega};$

g[{xnew_, vnew_}] :=
{x[T], v[T]} /. Flatten[NDSolve[{v'[t] == x[t] − x[t]^3 − γ v[t] + d Cos[ω t],
x'[t] == v[t], x[0] == xnew, v[0] == vnew}, {x, v}, {t, 0, T}]]
f[{x_, y_}] := {d, x}
totallyawesomegraph = bifurcation[.23, .38, 300, 150, 100];

Problem 10. Expand one or more of the parts of the bifurcation diagram that seem interesting.

13.4.2 Phase Space Trajectories of the Duffing Oscillator

With the bifurcation diagram of the Duffing oscillator in hand, it is possible to choose a value for *d* that apparently gives interesting behavior, such as period doubling. Sometimes it is desirable to see the phase-space trajectory for this particular value of the driving parameter. We now try to see how we can obtain this phase space plot, but it won't come easy.

> The following work in Mathematica is rather difficult. You may choose not to try to follow all of the details and instead move simply to the final function that graphs a trajectory in phase-space, namely **phasespaceorbit**.

We begin with

   In[57]:= **Clear["Global`*"]**

We define the parameters we wish to use.

   In[58]:= $\gamma = 0.1;\ \omega = 1.4;\ d = 0.318;\ T = \dfrac{2\pi}{\omega};$

Recall that when the function g, defined earlier, was used to iterate, we developed a set of initial points (x, v) for the beginning of each interval of one period in length for which we used **NDSolve**. We essentially threw the interpolating functions away. Here is that function again, which we shall need below.

   In[59]:= **g[{xnew_, vnew_}] :=**
       {x[T], v[T]} /. Flatten[NDSolve[{v'[t] == x[t] − x[t]³ − γ v[t] + d Cos[ω t],
       x'[t] == v[t], x[0] == xnew, v[0] == vnew}, {x, v}, {t, 0, T}]]

What we need now is a function that *plots* these interpolating functions, given these values of x and v as initial values. The function h, below, does this. It is almost the same as the function g, but it provides the entire solution of the differential equation for one period T of the driving force.

   In[60]:= **h[{xnew_, vnew_}] :=**
       Flatten[NDSolve[{v'[t] == x[t] − x[t]³ − γ v[t] + d Cos[ω t],
       x'[t] == v[t], x[0] == xnew, v[0] == vnew}, {x, v}, {t, 0, T}]]

The next function, **interpfunctions**, maps h over the set of initial values developed by the g function. Remember, we throw away **mtrans** early iterates and keep **nplot + 1** iterates.

   In[61]:= **interpfunctions[mtrans_, nplot_] :=**
       Map[h, NestList[g, Nest[g, {0, 0}, mtrans], nplot]]

This is how we apply **interpfunctions** to get a set of **nplot + 1** interpolating functions

   In[62]:= **plottable = interpfunctions[100, 50];**

Next, we put this table of these interpolating functions in the form that **ParametricPlot** can plot them all at once.

13.4 Return of the Duffing Oscillator

In[63]:= **functiontable[nplot_, s_] :=
Table[{x[s] /. plottable[[j]], v[s] /. plottable[[j]]}, {j, nplot + 1}]**

This is how we call the previous function:

In[64]:= **functiontable[50, s];**

Finally, we plot the phase-space trajectory.

In[65]:= **ParametricPlot[Evaluate[functiontable[50, s]], {s, 0, T}];**

OK, here is the relief you've been looking for, a single function that plots the trajectory of the system in phase-space. All of its innards have been described above, so you are spared any additional explanation. This function discards **mtrans** sections of the phase space trajectory, each **T** in length, and then graphs **nplot** sections, also **T** in length. Input to the function consists of the driving parameter d, the damping parameter γ, the frequency ω of the driving force, the number of sections you wish to discard as transient **mtrans**, and the number of sections you wish to plot **nplot**.

In[66]:= **phasespaceorbit[d_, γ_, ω_, mtrans_, nplot_] := Module$\left[\{T\}, T = \frac{2\pi}{\omega}\right.$;
g[{xnew_, vnew_}] := {x[T], v[T]} /.
Flatten[NDSolve[{v'[t] == x[t] − x[t]^3 − γ v[t] + d Cos[ω t],
x'[t] == v[t], x[0] == xnew, v[0] == vnew}, {x, v}, {t, 0, T}]];
h[{xnew_, vnew_}] :=
Flatten[NDSolve[{v'[t] == x[t] − x[t]^3 − γ v[t] + d Cos[ω t],
x'[t] == v[t], x[0] == xnew, v[0] == vnew}, {x, v}, {t, 0, T}]];
interpfunctions[m_, n_] := Map[h, NestList[g, Nest[g, {0, 0}, m], n]];
plottable = interpfunctions[mtrans, nplot];
functiontable[n_, s_] :=
Table[{x[s] /. plottable[[j]], v[s] /. plottable[[j]]}, {j, n + 1}];
ParametricPlot[Evaluate[functiontable[nplot, s]],
{s, 0, T}, AxesLabel −> {"x", "v"}, PlotRange −> All];]**

From the bifurcation diagram we suspect that a driving force of 0.318 might show an orbit of period 4. So, we plot the trajectory in phase space, throwing away 150 sections and keeping 50. We use our hard-won function **phasespaceorbit**.

In[67]:= **phasespaceorbit[.318, .1, 1.4, 150, 50]**

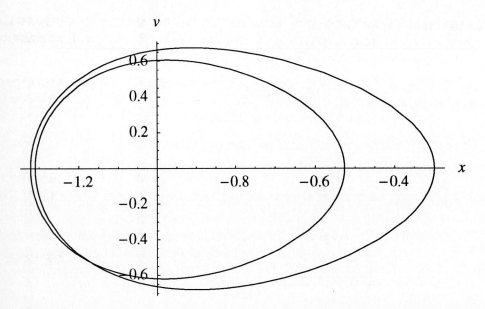

Wrong! It is only a period 2 orbit. The bifurcation diagram is deceptive (but not incorrect) at this point. Think of the Duffing oscillator as a particle oscillating around in a double well. What is happening near $d = 0.318$ is that the particle is oscillating around the well at $x = -1$ with period 2. Increasing or decreasing the driving force only slightly makes the particle jump to the other well where it also oscillates with period 2. This appears to be the case from $d \simeq 0.31$ to $d \simeq 0.333$ and may be true elsewhere.

Problem 11. Is the conjecture above true? Increase d to 0.319 and plot the phase-space trajectory. In which well, right or left, is the particle and what is its period of oscillation back and forth in this well? Also, decrease d to 0.316. Where is the particle and what is its period?

Problem 12. Can you find a value of d for which the particle oscillates with period 4?

13.4.3 Poincaré Sections

In[69]:= **Clear["Global`*"]**

Another way of studying the behavior of a system such as the Duffing oscillator is with the use of a *Poincaré section*. When we constructed the bifurcation diagram, we generated a extensive set of points (x_i, v_i) with the function g, defined and discussed in two previous sections.

We threw away the velocities and plotted the x_i versus d, the driving force. If instead, we plot each of the points (x_i, v_i) we obtain a Poincaré section, which is a sampling of the phase-space trajectory once each period of the driving force. Here is a function that will create a Poincaré section.

13.4 Return of the Duffing Oscillator

In[70]:= poincare[d_, γ_, ω_, mtrans_, nkeep_] := Module$\left[\{T\}, T = \frac{2\pi}{\omega};\right.$

g[{xnew_, vnew_}] := {x[T], v[T]} /.
Flatten[NDSolve[{v'[t] == x[t] − x[t]³ − γ v[t] + d Cos[ω t],
x'[t] == v[t], x[0] == xnew, v[0] == vnew}, {x, v}, {t, 0, T}]];
ListPlot[NestList[g, Nest[g, {0, 0}, mtrans], nkeep],
PlotStyle −> PointSize[.001], PlotRange −> All,
AxesLabel −> {TraditionalForm[x], TraditionalForm[v]}];]

We apply it to our situation

In[71]:= **poincare[0.35, .1, 1.4, 200, 5000]**

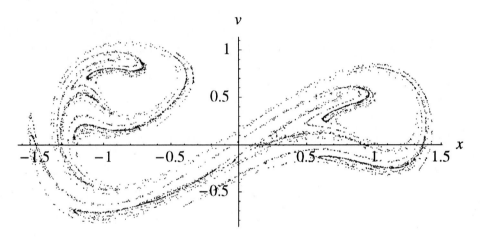

and get this weird kind of diagram that is called a *strange attractor*. (It would be wise not to graph 5,000 points on your first attempt. Work your way up to more points.)

Problem 13. Construct and interpret Poincaré sections for the situation γ = 0.1, ω = 1.4, d = 0.28 and also d = 0.318. Make **mtrans** = 200 and **nkeep** = 100. Why do you not get the myriad of points we did with d = 0.35? If the transients have disappeared, and the oscillator has period 1, how many points will there be on the Poincaré section? If the oscillator has period 2, how many points will there be on the Poincaré section? Why do we get thousands of different points in our figure?

There is another important property of the strange attractor. We illustrate with a somewhat more compact Poincaré section, which appears at the beginning of the chapter, and for which the parameters were

In[72]:= γ = 0.24; d = 0.68; ω = 1.7; T = $\frac{2\pi}{\omega}$;

We take the figure at the beginning of the chapter and expand the tail near the lower right-hand corner. Notice that the fine structure persists. If we were to plot many many more points and expand this figure, we would find still additional fine structure characteristic of the structure we see on the larger scales. It appears that no matter how detailed we look, we always see structure in this Poincaré section. That is also one of the characterstics of *fractals*. Is our strange attractor a fractal?

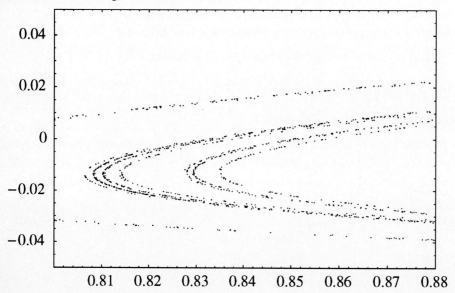

13.5 What Is Chaos?

In[73]:= **Clear["@"]**

We begin with what chaos is not. Contrary to a popular belief, chaos incorrectly implies that a system is not deterministic. If a system is not deterministic then you cannot predict the position x and the velocity v at any time subsequent to $t = 0$. The results are random. This is a false conception of what physicists mean by chaos. Here is a solution to the Duffing oscillator.

In[74]:= **γ = 0.1; ω = 1.4; d = 0.38;**

**solution = NDSolve[{v'[t] == x[t] − x[t]^3 − γ v[t] + d Cos[ω t],
    x'[t] == v[t], x[0] == 0, v[0] == 0},
    {x, v}, {t, 0, 100}, MaxSteps -> 2000] // Flatten**

13.5 What is Chaos?

In[75]:= **ParametricPlot[{x[s], v[s]} /. solution, {s, 0, 100}, AxesLabel -> {"x", "v"},
Compiled -> False];**

No matter how many times you repeat these instructions, you will always get the same result, not random results. If you do not believe it, try it. Differential equations *are* deterministic. In a recent *Newsweek* article a physicist turned priest appears to suggest that chaos is the point at which God selects which of several different possible outcomes will happen (*Newsweek*, July 20, 1998, p. 49). That may be true, but given a set of initial conditions, there is *only one* possible outcome of a differential equation, as we have seen.

The essential feature of chaotic systems is that tiny differences in initial conditions (or other conditions) lead to large differences in the state of the system. The now well-worn metaphor is that a butterfly flapping its wings in Florida may lead to a tornado in North Dakota. A non-chaotic system does not have this *sensitivity to initial conditions*, as we shall now show. Systems are unpredictable only because tiny unmeasurable differences in initial conditions lead to widely divergent trajectories of the solution. It's a practical problem not a theoretical problem. It also appears that chaotic systems are never periodic and their attractors have a fractal structure.

We now demonstrate the insensitivity to initial conditions of the Duffing oscillator in a non-chaotic mode and its sensitivity to initial conditions in a chaotic mode.

Here are some parameters for which the Duffing oscillator is not chaotic.

In[76]:= $\gamma = 0.1; \omega = 1.4; d = 0.1;$

We find one solution using **NDSolve** and $x = 0$, $v = 0$ for initial conditions.

In[77]:= **solution1 = NDSolve[{v'[t] == x[t] − x[t]^3 − γ v[t] + d Cos[ω t],
x'[t] == v[t], x[0] == 0, v[0] == 0}, {x, v}, {t, 0, 150},
MaxSteps -> 2000];**

We find another solution with slightly different initial conditions, namely $x = 0.0001$ and $v = 0$.

In[78]:= **solution2 = NDSolve[{v'[t] == x[t] − x[t]^3 − γ v[t] + d Cos[ω t],
x'[t] == v[t], x[0] == 0.0001, v[0] == 0}, {x, v}, {t, 0, 150},
MaxSteps -> 2000];**

Next we graph the logarithm (to the base 10) of the absolute value of the difference in the *x*'s of these two systems as a function of time. We see that the difference fluctuates but the average *tends exponentially toward zero with time* after the transient behavior is past.

In[79]:= Plot[Log[10, Abs[(x[s] /. solution1) − (x[s] /. solution2)]],
    {s, 0, 150}, PlotRange −> All, AxesLabel −>
        {TraditionalForm[t], TraditionalForm[Log[Abs[x$_1$ − x$_2$]]]}];

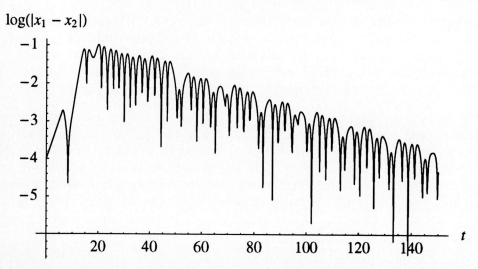

We repeat the experiment, but with a chaotic Duffing oscillator.

In[80]:= γ = 0.1; ω = 1.4; d = 0.38;

Once again we solve the differential equation given some initial conditions (same as above).

In[81]:= solution1 = NDSolve[{v'[t] == x[t] − x[t]^3 − γ v[t] + d Cos[ω t],
    x'[t] == v[t], x[0] == 0, v[0] == 0}, {x, v}, {t, 0, 100},
    MaxSteps −> 2000]

We solve again with slightly different initial conditions.

In[82]:= solution2 = NDSolve[{v'[t] == x[t] − x[t]^3 − γ v[t] + d Cos[ω t],
    x'[t] == v[t], x[0] == 0.00001, v[0] == 0}, {x, v}, {t, 0, 100},
    MaxSteps −> 2000]

Finally, we plot the logarithm of the difference in the *x*-values of the two systems as a function of time. Now we observe that average of the difference in the *x*-values *increases exponentially with time* until, of course, it reaches a limit because the particles can go only so high up the walls of the potential well, which you can see from the phase-space trajectory, the first graph in this section.

13.5 What Is Chaos?

```
In[83]:= Plot[Log[10, Abs[(x[s] /. solution1) - (x[s] /. solution2)]],
        {s, 0, 100}, PlotRange -> All, AxesLabel ->
          {TraditionalForm[t], TraditionalForm[Log[Abs[x₁ - x₂]]]}];
```

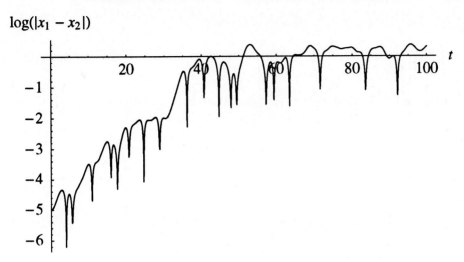

Now you know the essential difference in chaotic and nonchaotic systems, the key words being *sensitivity to initial conditions,* meaning that systems with slightly different initial conditions diverge *exponentially*.

Mostly Mathematica

1. Use **NestList** to find the first 100 iterates of the function $f(x) = \sqrt{x}$ starting with $x = 50$. Repeat this for a starting value of $x = 0.1$. What kind of point is $x = 1$? What is its basin of attraction?

2. Find these limits: $\lim_{n \to \infty} x^n/n!$, $\lim_{x \to 0} (e^x - 1)/x$, $\lim_{x \to 0} \sin(x)/x$.

3. Find the 100th iterate of $f(x) = \cos(x)$ without finding the first 99 iterates. Start with $x = 3$. Define

```
In[84]:= f[x_] = Cos[x]
```

Explorations

The very best exploration is to choose a system to study and study it; that is, plot some trajectories in phase-space, make a bifurcation diagram, make some Poincaré sections, and show that the system is sensitive to initial conditions. Here are some possible systems.

1. The damped and driven pendulum:

$$v'(t) = -\gamma v(t) - (1 + d\cos(\omega t))\sin(x(t)) \tag{4}$$

2. The spinning magnet (Briggs, 1987):

$$v'(t) = -\frac{M}{I} B_o \cos(\omega t) \sin(x(t)) \tag{5}$$

where x is the angle of the needle, M its magnetic moment, and B_o and ω are the amplitude and angular frequency of the varying magnetic field.

3. The various billard and bouncing ball problems described by Giordano (1997).

References

Briggs, K. Simple experiments in chaotic dynamics, *Am. J. Phys.* 55 (No. 12, December, 1987): 1083.

Giordano, N. J. *Computational Physics*, Upper Saddle River, N.J.: Prentice-Hall, 1997, p. 67.

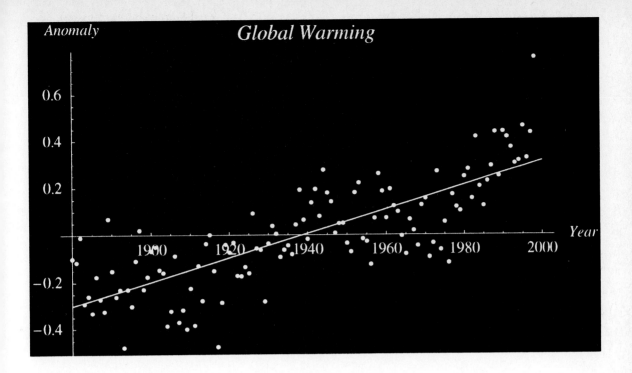

CHAPTER 14 Mathematica in the Laboratory

■ 14.1 Introduction

One of the tasks of the experimental physicist is to find a mathematical model that describes the results of an experiment. The model frequently takes the form of a mathematical relationship. For example, if we make a series of measurements of the voltage across a resistor as a function of the current through it, and if we make a graph of V versus I, the data will suggest a linear relationship between V and I. That is,

$$V \propto I \qquad (1)$$

or

$$V = RI \qquad (2)$$

which we know as Ohm's law. This is just one example of a *power-law* relationship, which is of the form

$$y = ax^b \qquad (3)$$

where a and b are parameters to be determined from the data. In particular, b is the exponent of the power law and may or may not be an integer. Of course, in the case of a *linear* relationship, $b = 1$, and in the case of an inverse relationship, b is negative. You should be able to think of other examples of power-law relationships: Here are two you will recognize: the first is between distance and time with constant acceleration,

$$s = \frac{1}{2} g t^2 \qquad (4)$$

and the second is between pressure and volume given constant temperature and a constant number of molecules of the gas,

$$PV = constant \qquad (5)$$

Of course, there may be more than one term in the function, for example

$$y = y_o + v_o t + \frac{1}{2} a t^2 \qquad (6)$$

in which case the quantity y is represented by a polynomial in t.

Problem 1. Think of at least two other power-law relationships with two different exponents. Make a rough sketch a graph of $s \propto t^2$. Also make a rough sketch of a graph of $P \propto 1/V$.

In cases where the relationship is not a power law, it is sometimes exponential. We might have a relationship of the form

$$y = ae^{bx} \qquad (7)$$

or

$$y = a(1 - e^{bx}) \qquad (8)$$

In this chapter we will see how Mathematica can help us decide which mathematical relationship best fits the data and how to determine the values of the parameters a and b.

Problem 2. Think of an exponential relationship between quantities in physics. Sketch graphs of Equation (7) and Equation (8). Assume $a = 1$ and $b = -1$.

Life is not always as simple as the relationships just outlined. In the case of an object falling in a resistive medium, we found one solution for the distance was

$$y(t) = \frac{v_T^2}{g} \ln\left(\cosh\left(\frac{g(t - t_o)}{v_T}\right)\right) \qquad (9)$$

14.2 Graphing and Analyzing Data

For most of us, it is not easy to anticipate the form of this curve. In any case, in this chapter we will look at graphing data and fitting theoretical models to the data.

■ 14.2 Graphing and Analyzing Data

In the next few sections we will look at a number of different examples of graphing and analyzing data. We begin with some old data on the recession velocities of galaxies obtained by Edwin Hubble in 1929 and compiled by Mays and Lesser (1998).

14.2.1 Something Old

In[1]:= **Clear["@"]**

To manage data we put them in a list. In this case the list consists of points whose first coordinate is the distance of a galaxy in Megaparsecs (Mpc) and whose second coordinate is the radial velocity of the galaxy in kilometers per second (1 parsec = $3.09 \cdot 10^{13}$ km). This is the data that led to the expansion of the universe concept. (A negative radial velocity means the galaxy is approaching.)

In[2]:= **velocitydata = {{0.032, 170}, {0.034, 290}, {0.214, −130}, {0.263, −70}, {0.275, −220}, {0.45, 200}, {0.5, 290}, {0.5, 270}, {0.63, 200}, {0.8, 300}, {0.9, −30}, {0.9, 650}, {.9, 500}, {1.0, 920}, {1.1, 450}, {1.1, 500}, {1.4, 500}, {1.7, 960}, {2.0, 850}, {2.0, 500}, {2.0, 800}, {2.0, 1090}}**

In[3]:= **g1 = ListPlot[velocitydata, AxesLabel −> {"Distance ", "Velocity "}, PlotStyle −> PointSize[.015]];**

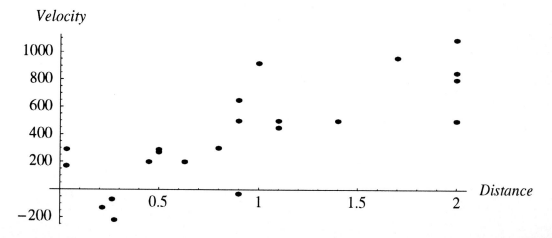

Perhaps you sense some correlation between distance and velocity, perhaps not. We proceed to fit a straight line to the data with Mathematica's **Fit** command.

In[4]:= **Fit[velocitydata, {1, x}, x]**

> In this case the **Fit** command will produce the constants m and b of a straight-line fit of the form $y = mx + b$.

We find the slope of the curve is 433 km Mpc^{-1} s^{-1} and the intercept is 0.67 km/s. Hubble massaged his data quite a bit and arrived at the result for the slope of 75 km Mpc^{-1} s^{-1}. Perhaps one thing he did was eliminate the negative velocity galaxies because clearly their velocity of motion in magnitude is larger than their velocity due to the expansion of the universe. The age we find for the universe is therefore

In[5]:= $\dfrac{10^6 \; 3.09 \; 10^{13}}{433} \; \dfrac{1}{3.16 \; 10^7}$

which is about 2 billion years. What are some current estimates of the age of the universe?

Problem 3. Hubble's value for the age of the universe was 13 billion years. Explain why the hypothesis of the Big Bang and the reciprocal of the slope of our model, converted to years, give the age of the universe. Simple kinematics arguments suffice, but be clear about your assumptions.

Let's plot the least squares fit curve on the same graph as the data.

In[6]:= **g2 = Plot[0.675 + 433 x, {x, 0, 2}, DisplayFunction −> Identity];
Show[g1, g2];**

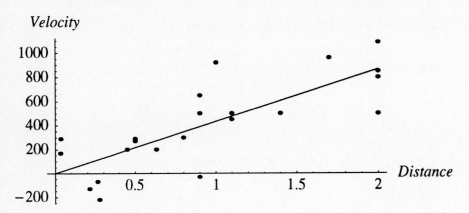

14.2 Graphing and Analyzing Data

> In case you have forgotten, the option **DisplayFunction -> Identity** allows the construction of the graph without its display. The **Show** command then displays both plots.

It would be nice to have some indication of the quality of the fit of the model. It is possible to fit a straight line to a set of points in x and y that are totally uncorrelated and arrive at a slope and an intercept, but they would be meaningless. There is a quantity, the *correlation coefficient*, which is used generate a measure of trust in the model. Here is a formula for its square:

$$r^2 = \frac{\sum_{i=1}^{n}(y_i^* - \bar{y})^2}{\sum_{i=1}^{n}(y_i - \bar{y})^2} \tag{10}$$

where $\bar{y}$ is the mean of the y-values of the data, y_i^* is the value of y from the model, and y_i is the actual value of y from the data. The best the model can fit is for the curve to go through every data point, in which case $r^2 = 1$ because in that case both the numerator and the denominator would be the same. A value of zero for r implies no correlation. Thus the hope is that we will obtain a value near 1 for r.

We proceed to calculate the correlation coefficient in steps. We begin by creating a function which takes a point in the dataset and returns the y- coordinate.

In[7]:= **k[{x_, y_}] := y**

This function is mapped over the data set to produce the y-values which are then added and divided by the number of elements in the data set to produce the mean, $\bar{y}$.

In[8]:= **ymean = Apply[Plus, Map[k, velocitydata]] / Length[velocitydata] // N**

The next function is our model with the best-fitting parameters entered. We will use it to find the numerator of r^2.

In[9]:= **y[x_] := 0.675 + 433 x**

Our next function, **h[{x_, v_}]**, finds a single term in the numerator of r^2.

In[10]:= **h[{x_, v_}] := (y[x] – ymean)^2**

The function **h[{x_, v_}]** is mapped over the data set and summed to give the numerator of the correlation coefficient.

In[11]:= **num = Apply[Plus, Map[h, velocitydata]] // N**

The function **j[{x_,y_}]** calculates a single term in the denominator of r^2.

In[12]:= **j[{x_, y_}] = (y − ymean)^2**

The terms in the denominator are summed,

In[13]:= **den = Apply[Plus, Map[j, velocitydata]] // N**

and r is calculated.

In[14]:= $\mathbf{rsquared = \pm\sqrt{\dfrac{num}{den}}}$

We obtained a value of 0.786, which, while not unity, does suggest a correlation between the distances and the velocities of the galaxies.

14.2.2 Something New

We obtained some data related to global warming from the National Climatic Data Center (NCDC) on the Internet (www.ncdc.noaa.gov). These data show the January through May (5-month) global land and ocean surface temperature anomalies for the years 1880-1988. The so-called anomaly is simply a deviation from some kind of *average*. Thus the actual temperature averaged over a 5-month period from January to May is the anomaly plus the *average*. Here is the data:

{{1880, −0.1}, {1881, −0.115}, {1882, −0.01}, {1883, −0.291}, {1884, −0.259},
{1885, −0.331}, {1886, −0.176}, {1887, −0.271}, {1888, −0.324}, {1889, 0.07},
{1890, −0.151}, {1891, −0.261}, {1892, −0.231}, {1893, −0.478},
{1894, −0.23}, {1895, −0.302}, {1896, −0.109}, {1897, 0.021}, {1898, −0.229},
{1899, −0.176}, {1900, −0.068}, {1901, −0.051}, {1902, −0.147},
{1903, −0.159}, {1904, −0.387}, {1905, −0.323}, {1906, −0.088},
{1907, −0.371}, {1908, −0.318}, {1909, −0.4}, {1910, −0.225}, {1911, −0.385},
{1912, −0.13}, {1913, −0.278}, {1914, −0.037}, {1915, 0.001}, {1916, −0.151},
{1917, −0.476}, {1918, −0.286}, {1919, −0.041}, {1920, −0.074},
{1921, −0.032}, {1922, −0.172}, {1923, −0.174}, {1924, −0.135},
{1925, −0.162}, {1926, 0.092}, {1927, −0.057}, {1928, −0.063},
{1929, −0.282}, {1930, −0.037}, {1931, 0.039}, {1932, 0.005}, {1933, −0.093},
{1934, −0.063}, {1935, −0.045}, {1936, −0.082}, {1937, 0.044}, {1938, 0.191},
{1939, 0.064}, {1940, −0.018}, {1941, 0.137}, {1942, 0.194}, {1943, 0.08},
{1944, 0.275}, {1945, 0.178}, {1946, 0.141}, {1947, 0.008}, {1948, 0.049},
{1949, 0.05}, {1950, −0.035}, {1951, −0.072}, {1952, 0.179}, {1953, 0.22},
{1954, −0.017}, {1955, −0.028}, {1956, −0.125}, {1957, 0.071}, {1958, 0.258},
{1959, 0.185}, {1960, 0.072}, {1961, 0.194}, {1962, 0.122}, {1963, 0.098},

14.2 Graphing and Analyzing Data

{1964, −0.004}, {1965, −0.081}, {1966, 0.066}, {1967, 0.019}, {1968, −0.046}, {1969, 0.126}, {1970, 0.154}, {1971, −0.094}, {1972, −0.035}, {1973, 0.268}, {1974, −0.063}, {1975, 0.055}, {1976, −0.118}, {1977, 0.171}, {1978, 0.118}, {1979, 0.102}, {1980, 0.246}, {1981, 0.278}, {1982, 0.154}, {1983, 0.416}, {1984, 0.205}, {1985, 0.123}, {1986, 0.227}, {1987, 0.292}, {1988, 0.437}, {1989, 0.249}, {1990, 0.439}, {1991, 0.415}, {1992, 0.371}, {1993, 0.303}, {1994, 0.314}, {1995, 0.462}, {1996, 0.325}, {1997, 0.433}, {1998, 0.754}}

Copy this data with an editing command and place it in an input cell under the name **tempdata**.

ListPlot is used to graph this data. Without doing a linear regression analysis, you can easily see what is happening to global temperatures on the surface of the Earth.

In[15]:= **g1 = ListPlot[tempdata, AxesLabel –> {"year", "ΔT (C)"}, PlotStyle –> PointSize[.015]];**

We do a linear regression analysis and plot the results.

In[16]:= **Fit[tempdata, {1, x}, x]**
g2 =
Plot[−9.9319 + .005123 x, {x, 1880, 2000}, DisplayFunction –> Identity];
Show[g1, g2];

You can see this graph at the head of the chapter.

Here is our calculation of the correlation coefficient, grouped into one cell, without additional explanation.

In[18]:= **k[{x_, y_}] := y**
ymean = Apply[Plus, Map[k, tempdata]] / Length[tempdata] // N;
y[x_] := −9.9319 + .005123 x
h[{x_, v_}] := (y[x] − ymean)^2
num = Apply[Plus, Map[h, tempdata]] // N;
j[{x_, y_}] := (y − ymean)^2
den = Apply[Plus, Map[j, tempdata]] // N;

$$\text{rsquared} = \pm\sqrt{\frac{\text{num}}{\text{den}}}$$

Our result for the correlation coefficient is 0.787. What do you think?

14.2.3 Linear Expansion of a Wire

In[23]:= **Clear["@"]**

This is a less dramatic example of expansion than that represented by the Hubble data: When heated, a wire increases its length in a linear fashion. Our data come from Trumper and Gelbman (1997). The following list gives a set of points whose first coordinate is the temperature (°C) and whose second coordinate is the length (mm):

In[24]:= **expansiondata = {{19, 812}, {39, 813.42}, {50, 813.79}, {64, 814.21},
{79, 814.29}, {102, 814.68}, {115, 814.76}, {129, 815.01},
{145, 815.10}, {166, 815.64}, {190, 815.74}, {215, 816.75}}**

Problem 4. Make a graph of this data and call it **g1**. Use Mathematica's **Fit** command to fit a straight line ($y = mx + b$) to the data. Make a graph of this line and call this graph **g2**. Show both **g1** and **g2** on the same coordinate system. From the slope of the line and the original length, 812 mm, estimate the expansion coefficient for this wire, which is made of Constantan. Compare this with the published expansion coefficient in a Handbook of Physics and Chemistry. Also find the correlation coefficient.

14.2.4 Old Law, New Data: Kepler's Third Law

In[25]:= **Clear["@"]**

Kepler's third law states

$$T^2 \propto a^3 \qquad (11)$$

where T is the period of the planet and a is its average distance from the Sun. Rephrased:

$$a \propto T^{\frac{2}{3}} \qquad (12)$$

Here is our first example of a power law with an exponent other than 1. Our list of data has nine points whose first coordinate is the period of revolution in millions of seconds and whose second coordinate is the average distance in 100 s of gigameters (10^{11} m).

In[26]:= **data = {{7.6, .579}, {19.4, 1.08}, {31.6, 1.50}, {59.4, 2.28}, {374., 7.78},
{935., 14.3}, {2640, 28.7}, {5220, 45.0}, {7820, 59.1}}**

We plot the data and note that it suggests a nonlinear relationship. You can also see one problem with a graph of this type: Because the data cover such a large range, some of the data are squashed near the origin and appear to give us little information.

In[27]:= **g1 = ListPlot[data, AxesLabel –> {"T", "a"}, PlotStyle –> PointSize[.015]];**

14.2 Graphing and Analyzing Data

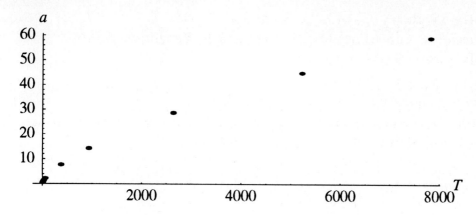

The cure for the squashing is also a cure for our problem of fitting a curve to the data. Instead of graphing the data, we graph the logarithm of the data. Mathematica will map the Log (natural log) over the data automatically, as is indicated by the following graph.

In[28]:= **g2 = ListPlot[Log[data], AxesLabel -> {"ln(T)", "ln(a)"},
         PlotStyle -> PointSize[.015]];**

This graph suggests a straight line, so we fit a line to the logarithmic data.

In[29]:= **Fit[Log[data], {1, x}, x]**

Notice that the slope is 2/3 as Kepler predicted!

We plot the line but suppress the graph.

In[30]:= **g3 = Plot[-1.89881 + 0.666761 x, {x, 1, 9}, DisplayFunction -> Identity];**

Finally, we superimpose the two graphs. (You have to wonder about that gap around ln(T) = 5; what's there?)

In[31]:= **Show[g2, g3];**

Here is our calculation of the correlation coefficient, grouped into one cell.

In[32]:= **k[{x_, y_}] := y**
ymean = Apply[Plus, Map[k, Log[data]]] / Length[data] // N
y[x_] := −1.89881 + .666761 x
h[{x_, v_}] := (y[x] − ymean)²
num = Apply[Plus, Map[h, Log[data]]] // N
j[{x_, y_}] = (y − ymean)²
den = Apply[Plus, Map[j, Log[data]]] // N

$$\text{rsquared} = \pm \sqrt{\frac{\text{num}}{\text{den}}}$$

Totally awesome! You did good, Kepler, even if you did not collect data of the quality we have now.

A power law relationship always gives a straight-line graph when the logs of the quantities are graphed – why? Suppose

$$s = t^n \qquad (13)$$

Then

$$\ln s = \ln t^n = n \ln t \qquad (14)$$

which is of the *form y = mx* with a slope of *n*. QED.

14.2.5 Errors, Errors, Errors

Mathematica's **Fit** function does not give us the estimated errors in the quantities it determines, but there is a standard package that does. We must first enter the package into memory with this command:

In[40]:= **<< Statistics`LinearRegression`**

> Use this command correctly the first time and don' try to execute **Regress** before loading the package.

Then we call the function **Regress**, which has the same form as **Fit**. Let's apply it to the Kepler's law data. We fit a straight line to the natural log of the data, hoping to show that the exponent of the power law is 2/3.

In[41]:= **Regress[Log[data], {1, x}, x]**

14.3 More Complex Data

That's a lot of output to assimilate. The important points are that the exponent is 0.666761 with a standard error (**SE**) of ±0.00024 and the **RSquared** number is 0.999999. In general, the closer **RSquared** is to 1, the better the fit. This is an extremely good fit indeed. Kepler's model works.

Problem 5. Apply this function to the linear expansion of the wire. What is the error in the slope and what is **RSquared**? Is the measured expansion coefficient of the wire within the SE of the published expansion coefficient?

Problem 6. Apply **Regress** to Hubble's data and interpret your results.

14.2.6 Motion of a Ball Thrown Upward

Next, we fit a polynomial to data we obtained from VideoPoint software, movie #DSONO17.MOV, made by the students at Dickinson College. We used VideoPoint to find the *y*-coordinate of a ball thrown into the air as a function of time. Then we cut the data from VideoPoint and pasted it into Mathematica in the form of two lists, the time data and the *y*-position data. The lists are joined and transposed to give the following time-position data for the ball:

> upball = {{0., 1.302}, {0.03333, 1.411}, {0.06667, 1.5}, {0.1, 1.578},
> {0.1333, 1.6456}, {0.1667, 1.703}, {0.2, 1.745}, {0.2333, 1.781},
> {0.2667, 1.807}, {0.3, 1.828}, {0.3333, 1.818}, {0.3667, 1.818},
> {0.4, 1.807}, {0.4333, 1.776}, {0.4667, 1.734}, {0.5, 1.682},
> {0.5333, 1.63}, {0.567, 1.552}, {0.6, 1.469}, {0.6333, 1.37},
> {0.667, 1.266}, {0.7, 1.151}, {0.733, 1.026}, {0.7667, 0.875},
> {0.8, 0.719}, {0.8333, 0.5573}, {0.867, 0.3854}, {0.9, 0.193},
> {0.9333, 0.0052}}

We plot the *y*-coordinate of the ball as a function of time.

> In[43]:= **ListPlot[upball, AxesLabel -> {"t (s)", "y (m)"},
>   PlotStyle -> PointSize[.015]];**

We fit a quadratic to the data. Why?

> In[44]:= **Fit[upball, {1, x, x^2}, x]**

From the kinematics of free-fall we know the coefficient of the squared term is $\frac{1}{2}g$, so we obtain $g = 9.94$ m/s/s.

Problem 7. Apply **Regress** to this data and find the **SE** of the quadratic term and the correlation coefficient. Is the actual value of g within the measured value of $g \pm$ **SE**?

14.2.7 Exponential Curves

Suppose a function decays exponentially. Thus

$$f = (A\,e)^{-Bx} \tag{15}$$

If we take the natural log of both sides of this equation we get

$$\ln f = \ln A - Bx \tag{16}$$

This equation is of the form $y = mx + b$ with $y = \ln f$, $b = \ln A$, and $m = -B$. For example, if A is 2 and B is π, then a graph of $\ln f$ versus x will yield a straight line with an intercept of $\ln 2$ and a slope of $-\pi$. With data such as this, make a new list with $\ln f$ as the first coordinate and x as the second. **ListPlot** this data to get a straight line. Apply **Fit[data, {1,x}, x]**.

Problem 8. Collect data that gives the voltage across a capacitor as a function of time as the capacitor discharges through a resistor. If you want to be able to read an ordinary voltmeter and do this, use a 1.0-farad capacitor and a resistance of approximately 100 ohms together with an ordinary stopwatch. With smaller components you will need some data recording instrumentation such as a computer connected to a ULI. Enter the data into Mathematica as a list of (t, v) points. Create a function to generate a new set of points $(t, \ln V)$. The graph of these points should be a straight line. Finally, fit a straight line to these points. The slope of the straight line should be $-1/RC$. This is a more or less standard lab in the second semester of a first-year physics course. Here is some simulated data with noise added if you don't have time to try the actual experiment.

{{0, 5.17911}, {2., 4.17904}, {4., 3.44676}, {6., 2.81531},
{8., 2.38264}, {10., 1.8552}, {12., 1.52421}, {14., 1.40983}, {16., 1.0423},
{18., 0.857766}, {20., 0.739966}, {22., 0.607173}, {24., 0.590128},
{26., 0.528867}, {28., 0.372861}, {30., 0.406529}, {32., 0.379575},
{34., 0.287408}, {36., 0.291385}, {38., 0.160952}, {40., 0.271687}}

■ 14.3 More Complex Data

In[46]:= **Clear["@"]**

14.3 More Complex Data

Many experimental problems can be handled using the techniques just described, including most power laws, polynomial functions, and exponential functions when there are only one or two parameters to determine. However, the world does not limit itself to these functions.

Greenwood, Hanna, and Milton (1986) have provided us with some time-distance data for a falling Styrofoam ball. This data, for a ball of radius 0.0254 m and mass 0.00254 kg, was obtained by videotaping, consequently the time intervals are a multiples of a 30th of a second. Here is the data arranged as a list of points with time in seconds as the first coordinate and position in meters as the second coordinate.

In[47]:= **fallingball = {{0, 0.075}, {3/30, 0.260}, {6/30, 0.525}, {9/30, 0.870},**
 {12/30, 1.27}, {15/30, 1.73}, {18/30, 2.23}, {21/30, 2.77}, {24/30, 3.35}}

We begin with a plot. The plot clearly shows that the relationship is not linear.

In[48]:= **g1 = ListPlot[fallingball, AxesLabel –> {"t", "y"},**
 PlotStyle –> PointSize[.015]];

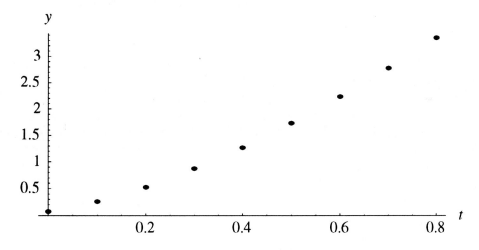

In Section 7.3.1, we solved the differential equation for an object falling under the influence of gravity and a resistive force proportional to the square of the speed. Here is the solution for the case where the object is dropped at $t = t_o$ at which time $y = 0$ and $v = 0$.

In[49]:= $y[t\_] = \dfrac{v_T^2}{g} \, \text{Log}\!\left[\text{Cosh}\!\left[\dfrac{g\,(t - t_o)}{v_T}\right]\right]$

In this equation v_T is the terminal velocity

$$v_T = \sqrt{\dfrac{mg}{k}} \qquad (17)$$

and *k* is the constant of proportionality between the resistive force and the square of the downward speed. We regard both v_T and t_o as parameters to be determined. We prefer not to work with subscripted variables in Mathematica, so we replace the two parameters with *a* and *b*. We also substitute in the value of *g*.

In[50]:= **y[t_] := $\frac{a^2}{g}$ Log[Cosh[g (t − b) / a]] /. g −> 9.8**

In the method of least squares the *x*-values (time in this experiment) are assumed to be accurate, while the measurement of the *y*-values (position in this experiment) is assumed to contain errors. For a given time, t_i, in the data list, we find the square of the difference between the theoretical model and the actual data. Here is a Mathematica function that does this.

In[51]:= **h[{t_, z_}] := (y[t] − z)^2**

The function **h[{t_, z_}]** takes a point **{t, z}** from the data list, plugs the **t** into the theoretical expression for **y** and subtracts the measured position **z**, and squares it. It forms, therefore, the square of the difference between the theoretical and the actual value of the position. We must map this function over each of the points in the list, and then we must add these squares. The following command does this.

In[52]:= **sumsquares = Apply[Plus, Map[h, fallingball]]**

Mathematica's **FindMinimum** function finds the value of *a* and *b* that minimizes the sums of the squares, but you must give it some initial guesses for each value. Here is the function, the parameters it is to find, and their initial guesses.

In[53]:= **FindMinimum[sumsquares, {a, 5.}, {b, .1}]**

The terminal velocity we found was 6.58 m/s. We make a graph of the theoretical model with the values found by **FindMinimum**.

In[54]:= **g2 = Plot[y[t] /. {a −> 6.58, g −> 9.8, b −> −.133}, {t, 0, 0.8},
        DisplayFunction −> Identity];**

To see how well the model fits, we graph the data and the model on the same coordinate system.

In[55]:= **Show[g1, g2];**

14.4 Importing Data

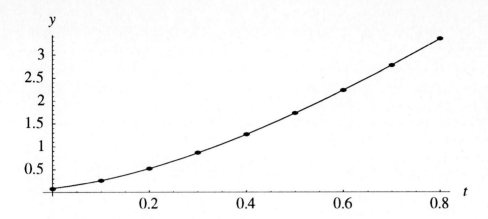

If you want to go first-class, then load the package **NonlinearFit** with this command.

In[56]:= << Statistics`NonlinearFit`

Apply the command **NonlinearRegress** to our **fallingball** data and the mathematical model we have chosen, specifying the parameters to be determined and starting values.

In[57]:= NonlinearRegress[fallingball,

$$\frac{a^2}{9.8} \text{Log}[\text{Cosh}[9.8\,(t-b)/a]], t, \{\{a, 5\}, \{b, 0\}\}, \text{ShowProgress} \rightarrow \text{True}]$$

is find a list of data for the falling shuttlecock measurements described in Chapter 3. The ordered pairs are in the form {time, distance}. Apply the same model that we used above and determine the terminal velocity and the starting time. Make a graph showing the data as well as the model. What is the terminal velocity?

In[58]:= shuttle = {{0.347, 0.61}, {0.47, 1.}, {0.519, 1.22}, {0.582, 1.52}, {0.65, 1.83},
{0.674, 2.}, {0.717, 2.13}, {0.766, 2.44}, {0.823, 2.74}, {0.87, 3.},
{1.031, 4.}, {1.193, 5.}, {1.354, 6.}, {1.501, 7.}, {1.726, 8.5}, {1.873, 9.5}}

Problem 10. Repeat Problem 9 but use **NonlinearRegress** from the statistics packages.

■ 14.4 Importing Data

In[59]:= Clear["@"]

In some instances, data is collected with a software package and some kind of laboratory interface. One way of transferring data from the software package to Mathematica is to **Cut** or **Copy** and **Paste** the data into Mathematica. Then use various list operations described in *The Mathematica Book* to manipulate the data into the kind of lists we worked with earlier in this chapter. Once you are familiar with list operations, it is relatively easy to transfer data quickly.

It is also possible to import data to Mathematica from text files. We have done this for files made with the Universal Lab Interface (ULI) and for files made with EXCEL. We will illustrate the process using a file we made with the ULI. The format of the data from the ULI is similar to a spreadsheet, a system of rows and columns. In this case we had four columns, so the appropriate form of **ReadList** is as follows:

In[60]:= **powerdata = ReadList["duracell", {Number, Number, Number, Number}];**

The name of the file was **duracell**, and we found that in order to read the file it had to be in the Mathematica folder on the hard disk. This was a long file, and if you don't want to see an entire file, you can see a much shorter version using the **Short** command, as follows:

In[61]:= **Short[powerdata, 24]**

> The **Short** command takes elements from *both* ends of the list, leaving all but 24 elements unprinted.

To find the actual length of the file we can use this command:

In[62]:= **Length[powerdata]**

It's time to explain what is in the file in order to understand what follows. The students were comparing D cells manufactured by various companies to see which one was the best. How do you decide what is best? It turns out this depend on the application of the battery. We arrived at the conclusion that the cell that provides the greatest energy per unit cost is best. How do you measure the electrical energy stored in a battery? We decided to attach the battery to a fixed resistor ($\approx 1.2\ \Omega$) and measure and monitor the power lost in the resistor ($I^2 R$) as a function of time. This data is in our **powerdata** list with time in the first column and power in the last column. (For our purposes, the middle two columns are of no concern here.) Thus the data are in the form

{{time 1, data, data, power 1}, {time 2, data, data, power 2}, ... ,{time n, data, data, power n}}

The **Extract** command can be used to extract elements from a nested list such as **powerdata**. Thus, the command

In[63]:= **Extract[powerdata, {{1, 1}, {1, 4}}]**

extracts the first and fourth element from the first set in the powerdata list. This next command sorts out all of the time and power data from the powerdata list. We show only a short form of the result.

14.4 Importing Data

In[64]:= **Short[Table[Extract[powerdata, {{i, 1}, {i, 4}}], {i, 1441}], 24]**

Finally, we plot the power versus time data from the file.

In[65]:= **ListPlot[Table[Extract[powerdata, {{i, 1}, {i, 4}}], {i, 1441}],**
   AxesLabel -> {"t", "P"}];

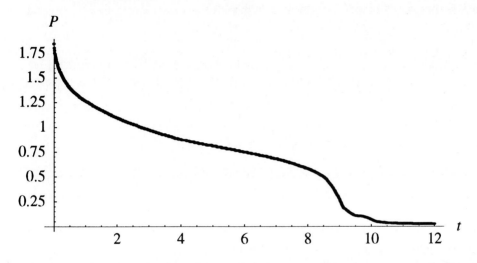

This is an interesting graph, and not significantly different from graphs for other alkaline batteries such as the Energizer. How do you find the energy from this graph? We know from the definition of power that

$$\text{Energy} = \int P(t)\,dt \qquad (18)$$

How can we integrate this function, since it is a table? We recall that an integral is basically a sum; thus

$$\int_a^b f(x)\,dx \simeq \sum_{i=1}^n f(x_i)\,\Delta x, \ \Delta x = \frac{b-a}{n} \qquad (19)$$

We can use a sum. In this case the data are spaced 0.01 h = 36 s apart, so Δt = 36 s. Another question is how far to integrate; from t = 0 to where? In the case of a flashlight application, when does the battery become useless? Is this different for a portable compact disk player application? We decided to stop the integration just past the shoulder of the curve where the power begins to drop rapidly. Specifically, we chose to stop when the power dropped below 0.5 W at t = 8.6 h or n = 8.6/0.01 = 860. Here is our integral:

In[66]:= **Sum[Extract[powerdata, {i, 4}] 36,**
   {i, 1, 860}]

Problem 11. What are the units of the answer? Make an argument that this answer is at least reasonable.

Mostly Mathematica

1. The parametric equations of the surface of a sphere are

$$x = r\sin(\phi)\cos(\alpha), \; y = r\sin(\phi)\sin(\alpha), \; r\cos(\phi) \qquad (20)$$

with α and ϕ parameters. Make a graph of a unit sphere using **ParametricPlot3D**. To begin, you may need to execute **??ParametricPlot3D**. Graph only the top half of the sphere. Then graph the bottom half of the sphere.

2. If **a** = {−1, 3, 5}, extract the *x*-component, the *y*-component, and the *z*-component of this vector. Here is a list: {*a, b, c, d, e*}. Extract the element *c* from this list.

3. Sum the elements in this list {$a, \pi, \sqrt{2}, x, y, z$} using **Apply**.

Explorations

1. Here is data relating the temperature of a cup of coffee (actually water) to time as the water in my favorite Einstein cup cooled toward the surrounding ambient temperature of the room, which was 20.5 °C. The data are in the form {time (minutes), temperature (°C)}. An alternative is to do your own cooling experiment, being more careful than I was to minimize cooling by evaporation and to keep the ambient temperature constant. Plot that data. Find a differential equation that models this situation, solve it with **DSolve**, and fit it to the data. Show both the data and the theoretical model on the same graph. Also find the correlation coefficient.

    In[67]:= **temp** = {{0, 61}, {1, 60}, {1.7, 59}, {2.9, 58}, {4, 57.5}, {5.6, 56.5},
        {6.8, 56}, {8.1, 55}, {10.1, 54}, {13.0, 53}, {15.6, 51.5}, {19.3, 50},
        {26.3, 47}, {32, 45}, {37.4, 43.5}, {43.6, 41.5}, {52.7, 39.25}, {62, 37},
        {68.7, 36}, {76.4, 34.5}, {88, 33}, {95, 32}, {109, 30.5}, {121, 29.5}}

2. Collect data obtained by measuring the voltage across a capacitor as it is charging through a resistor. Fit a model to this data to determine the product *RC*. Make the appropriate graphs and find the correlation coefficient.

3. Find and analyze other global temperature data. Some people claim global warming is not happening. Try to convince yourself one way or the other.

4. Using conductive paper used for plotting equipotential lines and electric fields, and with suitable electrodes painted on the paper and connected to a dc power supply, measure the voltage between the negative terminal and an array of points on the conductive paper.

14.4 Importing Data

Enter these into Mathematica as a list in the form (only an example of a 4-by-3 list of voltages) which follows:

In[68]:= **list = {{1.2, 1.3, 1.5, 1.6}, {1.3, 1.4, 2.1, 3.5}, {1.4, 2.2, 4.5, 5.7}}**

Then use **ListPlot3D** to make a plot of the potential as a function of two coordinates. Here is our graph of the potential of a conducting circular ring in the center of a rectangular conductor around the edge of the paper. You can see the potential is constant inside the ring, hence the electric field is zero in this region. This idea comes from Horn (1997), who used EXCEL to plot the data.

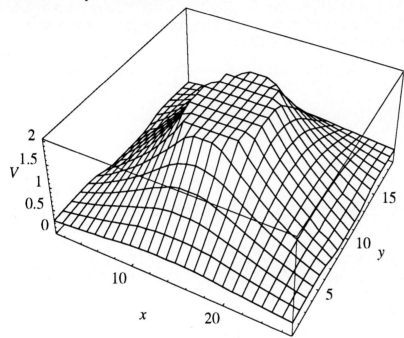

5. Here are some position versus time (meters, seconds) data for Donovan Bailey from the 1996 Summer Olympics (courtesy Wagner, 1998).

In[69]:= **{{0, 0}, {10, 1.9}, {20, 3.1}, {30, 4.1}, {40, 4.9}, {50, 5.6}, {60, 6.5}, {70, 7.2}, {80, 8.1}, {90, 9.0}, {100, 9.84}}**

Fit the model

$$f = v'(t) + \frac{v(t)}{\tau} \qquad (21)$$

to this data. First solve the differential equation with f and τ constants, then do a least-squares fit to determine the two constants. For additional details see Wagner (1998). Run your own 100-m dash to collect additional data.

References

Greenwood, M., C. Hanna, and J. Milton, Air resistance acting on a sphere, *The Physics Teacher*, (March 1986): 153.

Horn, J. Electrostatic landscapes, *The Physics Teacher* 35 (November 1997): 499.

Mays, R. L., and L. M. Lesser, *ACT in Algebra*, Boston, Mass.: WCB McGrawHill, 1998, p. 175.

Trumper R., and M. Gelbman, Measurement of a thermal expansion coefficient, *The Physics Teacher* 35 (No. 7, October 1997): 437.

Wagner, G. The 100-meter dash: Theory and experiment, *The Physics Teacher* 36 (March 1998): 144.

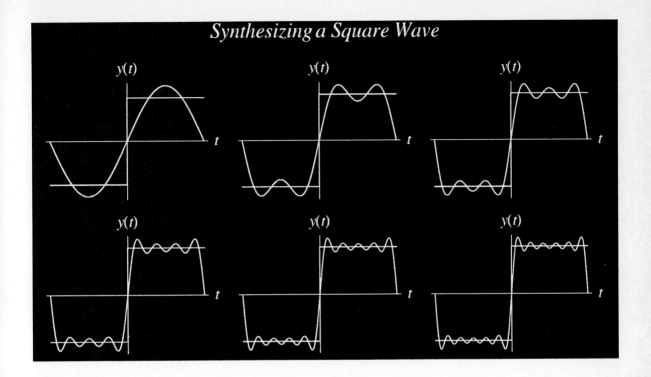
Synthesizing a Square Wave

CHAPTER 15 Additional Topics on Waves

■ 15.1 Introduction

It is possible to synthesize (almost) any periodic waveform with a harmonically related sum of sine and cosine waveforms. The figure above shows how sine waves may be added to *synthesize* a square wave. Equally important, it is possible to *analyze* an arbitrary waveform to see what frequencies are present. This is called spectrum analysis, and it takes place in almost every field you can imagine: physics, chemistry, engineering, music, medicine, biology, and forensics to name several. We begin the chapter by learning how to determine the amplitudes of the various components that make up the synthesized wave.

Next, we turn to the more practical problem of how to manage the same process when a waveform is only sampled at discrete points. This brings us to the so-called Fast Fourier Transform (FFT), which Mathematica has in its repertoire of commands.

We conclude the chapter with a brief look at a topic from quantum mechanics.

■ 15.2 Fourier Synthesis and Analysis

There is an important theorem by Fourier that is related to the superposition of waves and which has consequences both in experimentation and theory of wave phenemona. If $y(t)$ is a periodic function of time with period T, as it frequently is in the case of waves, then Fourier's theorem, which we accept without proof, says that $y(t)$ can be expressed as a sum of sine and cosine harmonics of a fundamental with frequency $f_o = 1/T$. Specifically

$$y(t) = \frac{1}{2} a_o + \sum_{n=1}^{\infty} [a_n \cos(2\pi n f_o t) + b_n \sin(2\pi n f_o t)] \qquad (1)$$

where

$$a_n = \frac{2}{T} \int_{-\frac{T}{2}}^{\frac{T}{2}} y(t) \cos(2\pi n f_o t) \, dt \qquad (2)$$

and

$$b_n = \frac{2}{T} \int_{-\frac{T}{2}}^{\frac{T}{2}} y(t) \sin(2\pi n f_o t) \, dt \qquad (3)$$

For the moment, do not try to gain a deep understanding of these equations, just trust them. At some later point in your physics career you may wish to spend some time thinking about why Fourier's theorem is true. We turn to a demonstration of these important results for several functions.

In[1]:= **Clear["Global`*"]**

We suppose $y(t)$ is the sawtooth wave of period 2.

In[2]:= **y[t_] := Mod[t – 1, 2] – 1**

In[3]:= **g1 = Plot[y[t], {t, –3, 3}, AxesLabel –> {"t", "y"}];**

15.2 Fourier Synthesis and Analysis

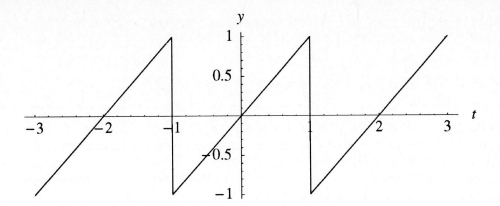

In this case, as we shall show below, the Fourier coefficients a_n are identically equal to zero, but the first 10 b-terms are

$$\{\frac{2}{\pi}, -\frac{1}{\pi}, \frac{2}{3\pi}, -\frac{1}{2\pi}, \frac{2}{5\pi}, -\frac{1}{3\pi}, \frac{2}{7\pi}, -\frac{1}{4\pi}, \frac{2}{9\pi}, -\frac{1}{5\pi}\}$$

Here is the synthesis of the sawtooth wave with 10 terms in the summation rather than an infinite number of terms

$$y(t) = \frac{2\sin(\pi t)}{\pi} - \frac{\sin(2\pi t)}{\pi} - \frac{\sin(4\pi t)}{2\pi} - \frac{\sin(6\pi t)}{3\pi} - \frac{\sin(8\pi t)}{4\pi} - \frac{\sin(10\pi t)}{5\pi} + \frac{2\sin(3\pi t)}{3\pi} + \frac{2\sin(5\pi t)}{5\pi} + \frac{2\sin(7\pi t)}{7\pi} + \frac{2\sin(9\pi t)}{9\pi} \quad (4)$$

and here is its graph.

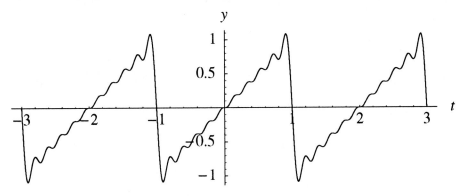

Finally, here is the original sawtooth wave and its synthesized counterpart superimposed.

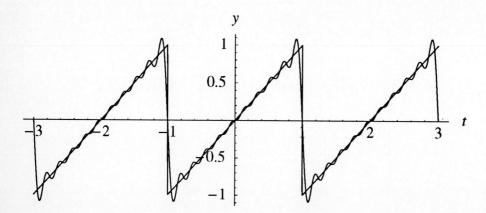

The synthesis proceeds as follows. Recall that the period of the fundamental in this case is 2; $T = 2$. Over the interval $t \in [-1, 1]$,

In[4]:= **y[t_] := t**

The Fourier coefficients are therefore

In[5]:= **T = 2;**

$$a[0] := \frac{2}{T} \int_{-\frac{T}{2}}^{\frac{T}{2}} y[t]\, dt$$

$$a[n\_] := a[n] = \frac{2}{T} \int_{-\frac{T}{2}}^{\frac{T}{2}} y[t]\, \text{Cos}\left[2\pi n\, \frac{1}{T}\, t\right] dt$$

$$b[n\_] := b[n] = \frac{2}{T} \int_{-\frac{T}{2}}^{\frac{T}{2}} y[t]\, \text{Sin}\left[2\pi n\, \frac{1}{T}\, t\right] dt$$

where we have made use of the fact that $f = 1/T$.

We can make tables of the Fourier coefficients with the **Table** command; thus

In[8]:= **Table[a[n], {n, 0, 10}]**

In[9]:= **Table[b[n], {n, 1, 10}]**

and we can graph the amplitude of the Fourier coefficients versus the frequency, $f = n/T$, to which they correspond.

15.2 Fourier Synthesis and Analysis

In[10]:= g1 = ListPlot[Table[{n $\frac{1}{T}$, Abs[b[n]]}, {n, 1, 25}],
    AxesLabel -> {"f", "b$_n$"}, PlotRange -> All,
    PlotStyle -> PointSize[.015], AxesOrigin -> {0, 0}];

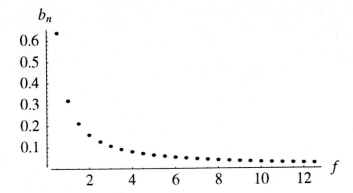

Here is the synthesis of y(t) for n terms

In[11]:= g[t_, n_] := $\frac{1}{2}$ a[0] + $\sum_{i=1}^{n}\left(a[i] \cos[2\pi i \frac{1}{T} t] + b[i] \sin[2\pi i \frac{1}{T} t]\right)$

Finally, we try finding **g[t,10]** to see if our expressions work

In[12]:= g[t, 10]

and we plot a few cycles with only five terms in the Fourier series:

In[13]:= Plot[Evaluate[g[t, 5]], {t, -3, 3}, PlotPoints -> 100,
    AxesLabel -> {"t", "y(t)"}];

Problem 1. Make a series of graphs, preferably a **GraphicsArray**, with successively larger numbers of terms in the Fourier series, say n = 1 to n = 6, for this sawtooth function.

Let's try to synthesize another periodic function of time, namely the square wave. We begin with a **Clear** because Mathematica has the coefficients for the sawtooth wave memorized. We use a **Global** clear because we have used a capital letter, **T**, in our calculations.

In[14]:= Clear["Global`*"]

Here is one way to describe one period of a square wave of amplitude one on the interval $t \in [-1/2, 1/2]$,

In[15]:= y[t_] := If[t >= 0, 1, -1]

and here is its graph over one period.

In[16]:= Plot[y[t], {t, -1/2, 1/2}, AxesLabel -> {"t", "y(t)"}];

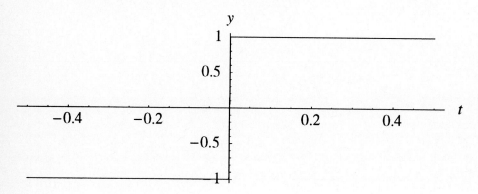

We compute the Fourier coefficients.

In[17]:= T = 1;

$$a[0] := \frac{2}{T} \int_{-\frac{T}{2}}^{\frac{T}{2}} y[t]\, dt$$

$$a[n\_] := a[n] = \frac{2}{T} \int_{-\frac{T}{2}}^{\frac{T}{2}} y[t] \cos\left[2\pi n \frac{1}{T} t\right] dt$$

$$b[n\_] := b[n] = \frac{2}{T} \int_{-\frac{T}{2}}^{\frac{T}{2}} y[t] \sin\left[2\pi n \frac{1}{T} t\right] dt$$

Problem 2. Make tables and graphs of a_n and b_n for n from 0 to 10. Note that there is no b_0 term.

Just as before, we write Fourier's theorem in Mathematica:

$$\text{In[20]:= } g[t\_, n\_] := \frac{1}{2} a[0] + \sum_{i=1}^{n} \left(a[i] \cos\left[2\pi i \frac{1}{T} t\right] + b[i] \sin\left[2\pi i \frac{1}{T} t\right] \right)$$

Problem 3. Plot the synthesized function on the interval $t \in [-1/2, 1/2]$ (one period) or a longer interval for a variety of n. Be sure to **Evaluate g** inside the **Plot** instruction or the result may take years to obtain.

Problem 4. Here is one cycle of a triangle wave; what is its period? Synthesize it; that is, find the Fourier coefficients, make tables of them, then plot the synthesized wave on the same axes as the wave itself. Are the a_n coefficients zero in this case? Make graphs of both a_n and b_n versus the frequency of the harmonic they represent.

In[21]:= **Clear["Global`*"]**

y[t_] := 1 − $\sqrt{t^2}$
Plot[y[t], {t, −1, 1}];

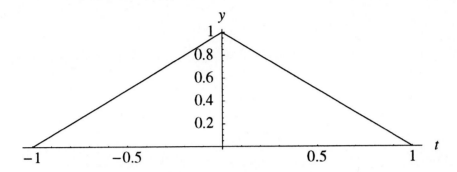

It is worth trying to come to some conclusion about the work we have just done. Given a time varying periodic signal y(t) (the note from a trumpet converted to a voltage by a microphone, for example) it is possible to find out what frequencies are present in that signal. Conversely, it is possible to synthesize a particular y(t) (a trumpet note emitted by a speaker which converts a voltage to a sound, for example) by combining the appropriate frequencies of sine and cosine waves. Decomposing a certain y(t) into sine and cosine terms is *spectrum analysis*. Combining sine and cosine terms of specific frequencies to produce a certain y(t) is *synthesis*.

■ 15.3 Discrete Fourier Analysis

In[24]:= **Clear["Global`*"]**

We wish to do a spectrum analysis of some time-varying function, f(t), which might be a result of you singing into a microphone. There are many complications, however, one of which is that in experimental work we do not have a continuous function f(t) to analyze. We always have a discrete set of points which are samples of f(t) at a given interval Δt. The integrals in Equations (2) and (3) must therefore be replaced by approximating sums. Mathematica has a command that produces a discrete Fourier analysis, but this introduces a second complication because it does a *complex* analysis; that is, it uses complex numbers.

We can illustrate the procedure simply. Suppose our sampled function gives us the points

In[25]:= **points = {1, 2, 3, 4}**

Here is Mathematica's discrete Fourier analysis, called a Fast Fourier Transform (FFT). (In Mathematica $\sqrt{-1} = \mathbf{I}$.)

In[26]:= **coefficients = Fourier[points]**

Notice that the output is a set of four complex numbers. We can combine these complex Fourier coefficients with Mathematica's **InverseFourier** command.

In[27]:= **InverseFourier[coefficients]**

There are apparently small imaginary terms, which we get rid of with **Chop,** a command that pitches out all terms less than 10^{-10}.

In[28]:= **Chop[InverseFourier[coefficients]]**

Lo and behold, we get back our original function.

To get a better feeling for what Mathematica's **Fourier** command does, let's assume our $f(t)$ is the periodic sawtooth wave we studied in the previous section. First, we need to sample it. FFTs usually work better and faster if the number of points in the sample is a power of 2: we choose $n = 256$. Using the following command, we will sample the interval $t \in [-1, 1]$, whose width is 2, at intervals of $\Delta t = 2/256$.

In[29]:= **Clear["Global`*"]**

In[30]:= **y[t_] = t;**

In[31]:= **n = 256; Δt = 2 / n;**

In[32]:= **sample = Table[N[y[t]], {t, −1 + 10^{-6}, 1, Δt}];**

> Sometimes we have to do tricky things to get what we want. Adding the 10^{-6} to the starting value of t gives us 256 points in the sample rather than the 257 we obtain if we start exactly at $t = -1$ and end exactly at $t = 1$.

We can check the length of the sample list this way.

In[33]:= **Length[sample]**

It is important to realize that we have just lost all of the time information in our sample of $y(t)$. To see a portion of the sample, execute this command:

In[34]:= **Short[sample, 10]**

15.3 Discrete Fourier Analysis

It is just a list of points. However, *we* know that they are separated by $\Delta t = 2/256$, so we haven't entirely lost the time information.

Now we do a FFT of this data.

In[35]:= **coefficients = Fourier[sample];**

We graph the *absolute value* of the complex coefficients.

In[36]:= **ListPlot[Abs[coefficients], PlotRange -> All, PlotStyle -> PointSize[0.01], AxesLabel -> {"n", "c$_n$"}];**

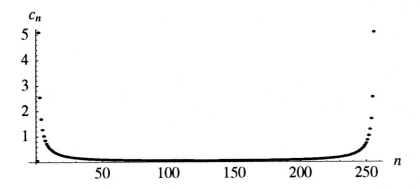

The symmetry in this graph illustrates a third complication. The coefficients to the right of $n = 128$ are redundant. Only half of the data shown is really meaningful. Also, because we lost the time data, we have also lost the frequency data, which we can get back. The frequency associated with each coefficient is $f_i = ((i-1)/(n\Delta t)$. We plan to make a graph of the absolute value of the coefficients for the first 25, just as we did in the last section. Before doing so, we note that the coefficients determined by Mathematica are calculated in a slightly different way than with Fourier's theorem for a continuous function, and in this case they are eight times larger. So, before plotting, we divide them by 8. Compare this graph with the corresponding graph in the previous section.

In[37]:= **g2 = ListPlot$\left[\text{Table}\left[\left\{\dfrac{i-1}{n\,\Delta t}, \text{Abs[coefficients[[i]]]}/8\right\}, \{i, 1, 25\}\right]\right.$,**
      AxesLabel -> {"f", "c"}, PlotRange -> All,
      PlotStyle -> PointSize[.015]$\Big]$;

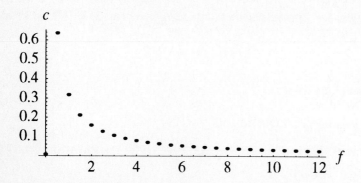

Finally, we reconstruct the waveform.

In[38]:= **ListPlot[InverseFourier[Chop[coefficients]]];**

However, this graph does not show the original time axis. We can get that back as follows:

In[39]:= **ListPlot[Table[{-1 + (i - 1) Δt,**
  InverseFourier[Chop[coefficients]][[i]]}, {i, 1, 256}],
  AxesLabel -> {"t", "y"}];

Perhaps this explanation has not been easy for you, so now we will provide a couple examples of spectrum analysis and you should find the process easier. In any case we are almost never interested in reconstructing the original function, but rather we want to know what frequencies are present and with what amplitude or power.

In[40]:= **Clear["Global`*"]**

Suppose our function consists of three sine waves of frequencies 13, 43, and 100 Hz with various phases.

In[41]:= $y[t\_] = \sin[2\pi\,13\,t] + \frac{1}{2}\cos\left[2\pi\,43\,t + \frac{\pi}{4}\right] + 2\sin\left[2\pi\,100\,t - \frac{\pi}{6}\right];$

To see what we're dealing with, let's plot our function.

In[42]:= **Plot[y[t], {t, 0, 1}];**

Let's sample the interval $t \in [0, 1]$ 256 times. This makes

In[43]:= **Δt = 1 / 256;**

We construct our sample, add some noise scattered about zero, check its length, and plot the sample.

In[44]:= **sample = Table[N[y[t]] + (Random[] − .5), {t, 0 + 10⁻⁶, 1, Δt}];
n = Length[sample]
ListPlot[sample];**

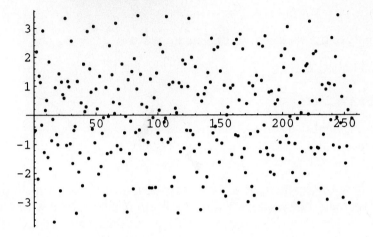

It is not clear there is anything periodic or systematic in this noisy looking signal. Now we find the FFT coefficients:

In[46]:= **coefficients = Fourier[sample];**

The spectrum is constructed next. First

In[47]:= $f := \dfrac{(i-1)}{n\,\Delta t};$

Then,

In[48]:= **ListPlot[Table[{f, Abs[coefficients[[i]]]}, {i, 1, 128}],
AxesLabel −> {"f(Hz) ", "Amplitude"}, PlotRange −> All,
PlotJoined −> True];**

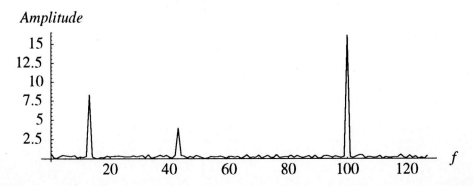

Behold, we found the original frequencies and in the appropriate amplitude ratios. In fact, divide the size of the coefficients by 8 and you have the original amplitudes of the trigonometric terms.

In[49]:= **Clear["Global`"]**

Problem 5. Use the FFT to analyze the spectrum of the square wave

In[50]:= **y[t_] = If[Mod[t, 1] > .5, 1, −1]**

over the interval $t \in [0, 10]$ with $n = 256$.

There is one extremely important qualification of this method that we need to state before concluding this section. Do not sample functions $y(t)$ with spectral components whose frequencies are larger than $f = 1/2\Delta t$. The method fails dramatically if you do.

Look at the figure at the beginning of Chapter 14; it represents deviations from some mean temperature for our planet for the past 120 years. Clearly the Earth appears to be warming. Are there any periodic trends in this data? We did an FFT on the data and obtained the following power spectrum, where the frequency is in years^{-1}. While there are some smaller peaks, none of them appear to be very significant. The large peak near 0.00085 y^{-1} represents a periodicity of about 120 years, but that is just approximately the interval over which the temperature has been rising. Is the temperature cyclic and are we about to see a downturn? Why might that be a foolish conjecture?

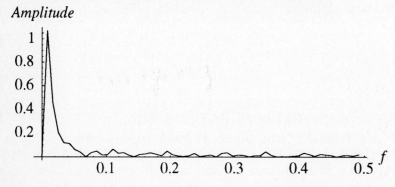

Fourier analysis is a powerful tool in many fields of study, and you now have a useful but preliminary understanding. Books have been written about Fourier series, transforms, FFTs, and applications, so at this point it would also be good to have some humility about your prowess with these techniques.

15.4 A Quick Look at Quantum Mechanics

Students in introductory physics courses sometimes hear a lot about quantum mechanics: The conduction of electrons in a solid can only be understood with quantum mechanics; specific heats can only be calculated with quantum mechanics; the behavior of molecules and atoms can only be explained with quantum mechanics. But what is quantum mechanics? It is likely you need to take a few courses in quantum mechanics to know what it is, but here is a tiny exposure to the subject.

The basic postulates of classical mechanics are Newton's three laws, the second law being the one we use most often to solve problems. Perhaps we should include the universal law of gravitation as one of the basic postulates. These laws appear to operate on a scale as small as people bumping into each other on a football field and also on a scale as large as the universe. They fail, however, to correctly describe atoms. At that scale and much smaller scales, the basic postulates of Newton are replaced by the *Schödinger equation*, which in one dimension for a single particle is:

$$i\hbar \frac{\partial \Psi}{\partial t} = -\frac{\hbar^2}{2m}\left(\frac{\partial^2 \Psi}{\partial x^2}\right) + V\Psi \qquad (5)$$

In this equation $\hbar$ is Planck's constant divided by 2π (it is called *h*-bar), m is the mass of the particle, V its potential energy function, and Ψ is called the *wave function*. Curiously, the equation contains an imaginary number in it. This equation neither fell from the sky nor came off the mountain with Moses. It is similar in form to the wave equation we studied in Chapter 10, and the wave equation and standing waves had a lot to do with the evolution of quantum mechanics in the minds of physicists. The really mysterious element in this equation is Ψ, for it does not represent a physical quantity, such as position or momentum, that we can measure.

However, $\Psi^*(x)\Psi(x)$ does mean something; it is proportional to the probability of finding the particle at position x (Ψ^* is the complex conjugate of Ψ). The wavefunction Ψ is sometimes called the *probability amplitude*, and it must have a number of mathematical properties, including being single-valued and continuous with continuous derivatives. Perhaps what you want to know is whether or not Schrödinger's equation is true. Just as in the case of Newton's laws, nobody knows; it is a *postulate*. What is known is that this postulate has been enormously successful in predicting the results of experiments; indeed, physicists construct models, not truth.

If the potential energy of the particle does not depend on time, then the potential energy is a function of position only. In that case the following simplification takes place:

$$\frac{d^2\Psi}{dx^2} + \frac{2m}{\hbar^2}(E - V)\Psi = 0 \qquad (6)$$

which is the *steady-state* form of Schrödinger's equation (in one dimension). In this equation E is the energy of the particle.

We will consider a famous problem called *the particle in a box*. We imagine the particle to be in a one-dimensional box of length L, extending from $x = 0$ to $x = L$. Inside the box the potential energy is zero; at the extremely hard walls of the box the potential energy jumps to infinity. Thus the particle has to move inside the box. The particle cannot have infinite energy, so it cannot exist outside the box, and since the probability of finding it outside the box must be zero, the wave function is zero if $x \leq 0$ or $x \geq L$. Since $V = 0$ in the box, our differential equation becomes

In[51]:= **schrodinger = Ψ''[x] + $\dfrac{2m}{\hbar}$ E Ψ[x] == 0**

We attempt to solve it with Mathematica's **DSolve**.

In[52]:= **DSolve[schrodinger, Ψ[x], x]**

That's pretty simple, the solution is either a sine function or a cosine function or both, depending on what we find out for the constants of integration (**C[1]** and **C[2]**). We know that $\Psi(0) = 0$, so perhaps we should put that in for an initial condition. We try again.

In[53]:= **DSolve[{schrodinger, Ψ[0] == 0}, Ψ[x], x]**

We also know that $\Psi(L) = 0$, and one way to make sure of that is to also choose **C[1] = 0**. In this case, however, the wave function is zero everywhere in the box. That is, the probability of finding the particle in the box is zero, but this contradicts our hypothesis that there is a particle in the box. Another possibility is to choose our parameters so that the sine function in the solution is zero when $x = L$. This will be true if

$$\sqrt{\dfrac{2mE}{\hbar}}\, L = n\pi \tag{7}$$

where n is a nonnegative integer. We have arrived at the result that for the particle to be in the box it can only have certain energies. THE ENERGY IS QUANTIZED! The system has *energy levels*. Here are the possible energies from Equation (7):

$$E_n = \dfrac{n^2 \pi^2}{2mL^2}, \; n = 1, 2, 3, \ldots \tag{8}$$

The whole number n is called the *quantum number*. Not only can it not be negative, it cannot be zero because in that case the wave function is zero everywhere in the box, and the particle would not be in the box, contrary to our hypothesis that it is. Thus a particle in a box cannot have zero energy, even at a temperature of absolute zero, nor can it have just any other energy; it must have the energy given by Equation 8. This is in dramatic contrast

15.4 A Quick Look at Quantum Mechanics

to systems described by Newton's postulates in which a particle can have any energy, including zero.

Problem 6. Compute and graph the energy levels for an electron in a box with $L = 10^{-10}$ m. What is the lowest possible energy for the electron in electron volts? What is the speed of the electron in its lowest energy level?

Problem 7. Compute and graph the energy levels for a marble of mass 0.01 kg in a box with $L = 0.1$ m. What is the speed of the marble in its lowest energy level? Could you discern movement of a marble with this speed? Suppose the same marble has a speed of a few centimeters per second; what is its energy? In what energy level, n, is it? How fast would it be moving at the next higher energy level? Could you distinguish these two energies in the speed of the marble?

We can now also find the wave function. Substituting Equation (8) into the solution to the differential equation gives

In[54]:= $\Psi[x\_] := C[1] \, \text{Sin}\left[\dfrac{n \, \pi \, x}{L}\right]$

Since $\Psi(x)$ is real and not complex, the probability of finding the particle at position x is $(\Psi(x))^2$. Since it is *certain* (probability = 1) that the particle is in the box, the following integral is 1. Thus,

In[55]:= $\displaystyle\int_0^L (\Psi[x])^2 \, dx == 1$

We solve this equation for the integration constant **C[1]**.

In[56]:= **Solve[%, C[1]]**

Choosing the positive solution and noting that the sine term is always zero because n is an integer, we find

$\Psi[x\_] := \sqrt{\dfrac{2}{L}} \, \text{Sin}\left[\dfrac{n \, \pi \, x}{L}\right]$

for the wave function. We are now in a position to graph the wave functions. Choosing $L = 10$, we graph the wave functions.

In[58]:= **Clear["@"]**

In[59]:= **g1 = GraphicsArray[Table[{**
    Plot[Ψ[x] /. L -> 10, {x, 0, 10},
    DisplayFunction -> Identity, AxesLabel -> {"x", "Ψ"}],

```
     Plot[Ψ[x]² /. L -> 10, {x, 0, 10},
     DisplayFunction -> Identity, AxesLabel -> {"x", "Ψ²"}]}, {n, 1, 4}]];
     Show[g1];
```

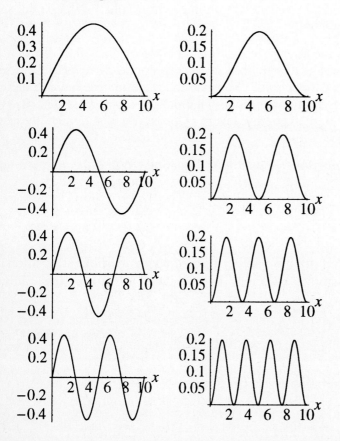

The first column of graphs show $\Psi(x)$, while the second column shows $\Psi^*(x)\Psi(x)$.

At the lowest energy, the most likely place for the particle is in the center of the box; at higher energies there are forbidden places in the box. In classical mechanics a marble in a box can be anywhere in the box with equal likelihood. Certainly quantum mechanics is significantly different than the mechanics of Newton's laws, which is the mechanics of bicycles, skyscrapers, 747s, planets, stars, and galaxies. Quantum mechanics must be used when dealing either individually or collectively with subnuclear particles, nuclei, atoms, and molecules.

If you put the particle in a two-dimensional box with dimensions L and M, you get two sets of quantum numbers. Without showing the proof, the wave functions become

In[60]:= **Clear["Global`*"]**

In[61]:= $\Psi[x\_, y\_] := \sqrt{\dfrac{4}{LM}} \; \text{Sin}\left[\dfrac{n\pi x}{L}\right] \text{Sin}\left[\dfrac{m\pi y}{M}\right]$

We make a density plot of $\Psi^*\Psi$ to give a nice picture of the probability of finding the particle at a particular position. We choose $L = 2$ and $M = 1$ to get a rectangular box and we pick quantum numbers $n = 7$, $m = 2$.

In[62]:=

DensityPlot[(Ψ[x, y] /. {L -> 2, M -> 1, n -> 7, m -> 2 })^2, {x, 0, 2}, {y, 0, 1}, Mesh -> False, PlotPoints -> 300];

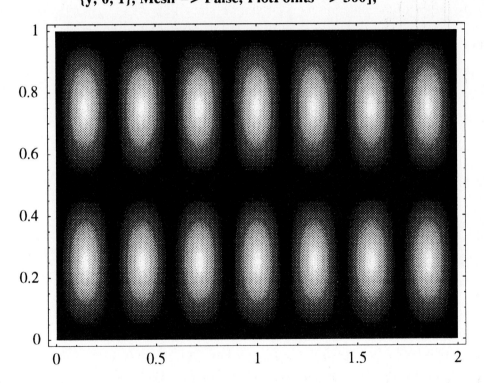

Problem 8. If we could watch without disturbing an electron (you can't), in a two-dimensional box with quantum numbers $n = 7$ and $m = 2$ it might appear to hang around in one of the probability density maxima, and then more-or-less jump from one probability density maxima to the next. You can see that from the density plot. What happens to the density plot for large quantum numbers? Try $n = 50$ and $m = 50$.

Problem 9. Make a **Plot3D** plot of the probability density function $\Psi^*\Psi$ for the same rectangular box.

We would like to go on. The next problem might be a particle in a one-dimensional quadratic potential well, $V(x) = kx^2/2$, the harmonic oscillator. Then we would move to the hydrogen atom, which we explore in a little more detail below. However, even in a problem of such apparent simplicity the mathematics becomes formidable, and approximate methods such as numerical integration are used. The hydrogen atom, molecules, and the quantum mechanics of solids, liquids, and gases are well beyond the scope of this book. The simplicity of the differential equation that is the basic postulate of quantum mechanics is deceptive.

Mostly Mathematica

1. Do an FFT of this set of data: {0, 1, 2, 3, 4, 5, 6,7}. Then do an inverse FFT and **Chop** to get the original function back again.

2. Make a **DensityPlot** of the function $f(x, y) = \sin(x + \sin(y))$ for $x \in [-3, 3]$ and $y \in [-3, 3]$. Once you have done this with an option of **PlotPoints -> 100**, add the option **ColorFunction -> Hue**. (From Gray and Glynn, 1991).

3. Make contour maps and density maps of the function

$$\Psi[x, y] := \sqrt{\frac{4}{LM}} \sin\left(\frac{n \pi x}{L}\right) \sin\left(\frac{m \pi y}{M}\right) \qquad (9)$$

for $L = 2$, $M = 1$, $n = 6$, $m = 3$. Use the options **ColorFunction -> Hue**, **Mesh -> False**, and **PlotPoints -> 200**. Let $x \in [0, 2]$, $y \in [0, 1]$.

Explorations

1. Do a Fourier analsis of a number of different functions on the interval $[-1, 1]$. In particular, look at some functions that are even and some that are odd. An even function is one for which

$$f(-x) = f(x) \qquad (10)$$

while an odd function is one for which

$$f(-x) = -f(x) \qquad (11)$$

Functions such as $\cos(x)$, x^2 and x^4 are even, while examples of odd functions include $\sin(x)$, x, and x^3. What Fourier coefficients are zero if the function is even? Odd?

2. A limitation we placed on discrete Fourier analysis was to exclude from analysis those waveforms that had frequencies larger than $f = 1/2\Delta t$ where Δt was the sampling interval. What happens if you disobey that rule? Explore the consequences starting with a waveform that has frequency components at $f = 1/2\Delta t$ and at $f = 1/\Delta t$.

3. Convince your instructor to either create an experiment involving Fourier analysis or demonstrate Fourier analysis in the laboratory. You might analyze the notes from a variety of musical instruments. How does a note from a flute compare in complexity to a note

15.4 A Quick Look at Quantum Mechanics

from an oboe?

4. Apply Fourier analysis techniques to the damped and driven oscillations of the Duffing (or other) oscillator for various values of the driving force. You can analyze either $x(t)$ or $v(t)$. What happens to the spectrum when the oscillator moves from period 1, to period 2, to chaotic behavior? Also analyze the time variation of the butterfly population both when it is chaotic and when it is nonchaotic (see Section 9.3).

5. In solving Schrödinger's equation for the hydrogen atom, we begin by setting up a spherical coordinate system r, φ, θ, where φ is the azmiuthal angle measured in the x-y plane around the z-axis and θ is measured from the z-axis down to the **r**-vector. You may need to do some research to understand this coordinate system and the other details that follow. The variables can be separated and the solutions are

$$\text{In[63]}:= \ \Phi[m\_, \varphi\_] := \frac{1}{\sqrt{2\pi}} e^{I m \varphi}$$

$$\text{In[64]}:= \ \Theta[l\_, m\_, \theta\_] := \sqrt{\frac{(2l+1)(l-\text{Abs}[m])!}{2(l+\text{Abs}[m])!}} \ \text{LegendreP}[l, \text{Abs}[m], \text{Cos}[\theta]]$$

$$\text{In[65]}:= \ R[n\_, l\_, r\_] :=$$
$$-\sqrt{\left(\frac{2}{n a}\right)^3 \frac{(n-l-1)!}{2n(n+l)!}} \ e^{-\frac{r}{na}} \left(\frac{r}{a}\right)^l \text{LaguerreL}\left[n-l-1, 2l+1, \frac{r}{a}\right]$$

where

$$\text{In[66]}:= \ a = \frac{\hbar^2}{\mu e^2}$$

(You will need to refer to a quantum mechanics or modern physics textbook for any additional explanation of this solution.)

In this last equation, $\hbar$ is Planck's constant divided by 2π, μ is the reduced mass of the hydrogen atom and is very close to the mass of the proton, and e is the charge of the proton. In the earlier expressions r, φ, and θ are the coordinates, and n, l, and m are the quantum numbers analogous to those we found for the particle in a box. Here is the wave function for the hydrogen atom.

$$\text{In[67]}:= \ \Psi[r\_, \theta\_, \varphi\_, n\_, l\_, m\_] := \Phi[m, \varphi] \, \Theta[l, m, \theta] \, R[n, l, r]$$

We will use **DensityPlot** to visualize the probability density in two dimensions. We set $\varphi = 0$, which means we will be viewing a cross section of the probability density with the z-axis in the plane of the paper and oriented up-and-down. We illustrate with the quantum numbers $n = 3$, $l = 1$, and $m = 0$. Since $\varphi = 0$, $y = 0$, and we are working in the x-z plane. In that case

In[68]:= **r = √ x^2 + z^2**

and

In[69]:= **θ = ArcTan[x / z]**

We choose a set of units in which

In[70]:= **a = 1;**

Here is the plot.

In[71]:= **DensityPlot[Ψ[r, θ, 0, 3, 1, 0] ^ 2, {x, −25, 25}, {z, −25, 25}, PlotPoints −> 200, Mesh −> False];**

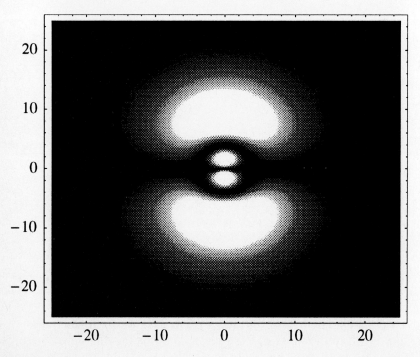

Use **DensityPlot** to explore some of the various probability density distributions in more detail, but not without researching the meaning of the various quantum numbers and the solutions to Shrödinger's equation for the hydrogen atom.

Reference

Gray, T. W. and J. Glynn, *Exploring Mathematics with Mathematica*, Reading, Mass.: Addison-Wesley 1991, p. 182.

UNIVERSITY OF ST. THOMAS LIBRARIES
WITHDRAWN
UST
Libraries

QC 20 .D33 1999
De Jong, Marvin L.
Mathematica for calculus-
 based physics

DATE DUE

Demco, Inc. 38-293